CHEMICAL EQUILIBRIUM

A PRACTICAL INTRODUCTION FOR THE PHYSICAL AND LIFE SCIENCES

CHEMICAL EQUILIBRIUM

A PRACTICAL INTRODUCTION FOR THE PHYSICAL AND LIFE SCIENCES

William B. Guenther

Department of Chemistry
The University of the South
Sewanee, Tennessee

PLENUM PRESS · NEW YORK AND LONDON

Library of Congress Cataloging in Publication Data
Guenther, William B
Chemical equilibrium.

Bibliography: p.
Includes index.
1. Chemical equilibrium. I. Title.
QD503.G8 541'.392 75-28028
ISBN 0-306-30850-9

A Division of Plenum Publishing Corporation
227 West 17th Street, New York, N. Y. 10011

United Kingdom edition published by Plenum Press, London
A Division of Plenum Publishing Company, Ltd.
Davis House (4th Floor), 8 Scrubs Lane, Harlesden, London, NW10 6SE, England

Printed in the United States of America

* Preface

The present work is designed to provide a practical introduction to aqueous equilibrium phenomena for both students and research workers in chemistry, biochemistry, geochemistry, and interdisciplinary environmental fields. The pedagogical strategy I have adopted makes heavy use of detailed examples of problem solving from real cases arising both in laboratory research and in the study of systems occurring in nature. The procedure starts with mathematically complete equations that will provide valid solutions of equilibrium problems, instead of the traditional approach through approximate concentrations and idealized, infinite-dilution assumptions. There is repeated emphasis on the use of corrected, conditional equilibrium constants and on the checking of numerical results by substitution in complete equations and/or against graphs of species distributions. Graphical methods of calculation and display are used extensively because of their value in clarifying equilibria and in leading one quickly to valid numerical approximations.

The coverage of solution equilibrium phenomena is not, however, exhaustively comprehensive. Rather, I have chosen to offer fundamental and rigorous examinations of homogeneous step-equilibria and their interactions with solubility and redox equilibria. Many examples are worked out in detail to demonstrate the use of equilibrium calculations and diagrams in various fields of investigation. Over 100 other exercises are included, most with answers and hints for solution. I have tried to bridge the gap between the oversimplified treatment in general and analytical chemistry texts and the complexities of the advanced presentations of multiple competing equilibria in real systems found in the books by Butler, Garrels and Christ, and Stumm and Morgan (see the Bibliography, Appendix A-6). Students and workers in modern biology and geology as well as in chemistry need more help than these works offer in using equilibrium calculations to deal correctly with important aqueous systems. In this emphasis, this book differs from other works in that they are directed more toward chemical analysis, often giving scant coverage to the earth and biosciences and to applications to nonideal solutions.

The present book differs also in its inclusion of varied pH diagrams, graphical solution methods, details of Gran plots, and the construction of E–pH diagrams. Descriptions of Gran (linear titration) plots in the literature often do not make clear their limitations in precisely those cases in which we want most to use them: extremes at which the approximate equations used to derive the linear functions do not apply. Finally, it is my hope that the gradual introduction of the complexities affecting solubility found in Chapter 10 will prove helpful in dealing with this difficult topic.

Throughout, I have attempted to adhere closely to standard IUPAC symbolism, as exemplified in the volumes of *Stability Constants* (see the Bibliography, Appendix A-6). The chief exception is in the use of acidity constants rather than the proton formation constants of anions; I do this because it seems still to be predominant usage, though I hope this will change.

I realize that many who are familiar with equilibrium calculations will find the complete equations, the log ratio, $\log \alpha$, $\log C$, and $\bar{n}$ diagrammatic methods new and strange. Nonetheless, I hope that they will persist in looking through this study, to discover how great are the gains in clarity and assurance when using these methods in preference to the guesswork of more traditional approaches.

The preparation of the large number of α, $\log \alpha$, and $\bar{n}$ diagrams was facilitated by use of an X–Y plotter and a computer program compiled by Dr. C. C. Ross, Jr., Mathematics Department, Director of Academic Computing, University of the South (see Appendices A-2–A-4). It seemed best not to retouch the small discontinuities, which arise from tolerances in the computation program and plotter in these plots.

I record my gratitude for the encouragement of many colleagues and for the inspiration of the writers of the books listed in the Bibliography, Appendix A-6. I thank the University of the

South for according to me a sabbatical in 1974–75 for the completion of the book.

William B. Guenther

Sewanee, Tennessee
August 1975

Contents

1 Electronic Structure in Aqueous Acid–Base Chemistry

1. The Proton in Compounds

Modern theory supports the view that chemical change in matter is the process of rearrangement of negative electron clouds and positive atomic kernels to form a more stable mutual relation. The lightest element, hydrogen, is unique. All others retain a core of inner electrons throughout their reactions. But the core of hydrogen is only a proton. It is about 10^{-5} the diameter of the lithium ion Li^+. Thus, the special ion H^+ has exceptionally high positive charge density, the charge per unit volume. It associates with electron clouds in any matter available. Other ions, like the Li^+, can exist as entities in some crystal lattices, while no substances contain H^+ as a separate entity or ion unit. It is important to be clear from the start about the vastly different species formed by hydrogen (Figure 1-1). When we speak of hydrogen ion and hydrogen compounds, we shall almost always mean the proton contained in an electron pair cloud as shown in the H_3O^+ scheme in Figure 1-1. For a full discussion of evidence and the unique features of proton chemistry see Chapter 5 in the book by Bockris and Reddy.[1]

In water, compounds having acidic protons (Brønsted acids) form hydronium H_3O^+ and higher hydrates, in which the proton is embedded in an unshared electron cloud on a water oxygen. Under

[1] J. O. Bockris and A. K. N. Reddy, *Modern Electrochemistry*, Plenum Press, New York, 1970, Vol. 1, Chapter 5, "Protons in Solution."

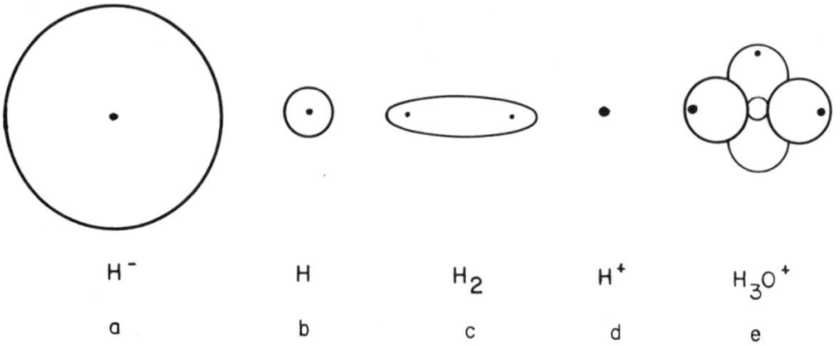

Figure 1-1. Drawings to scale of the forms of hydrogen. (a) The hydride ion, a proton with two electrons. (b) The hydrogen atom, a proton with one electron. (c) The hydrogen molecule, two separated protons in a cloud of two electrons. (d) The hydrogen core, or H^+ ion, a proton. (e) The hydronium ion, oxygen with four electron cloud pairs, three of which are protonated in H_3O^+.

common conditions only the first row elements from B to F, and a few others like S, can hold protons in water. In addition, the three smallest, N, O, and F, exhibit the phenomenon of "hydrogen bonding," which accounts for the unusual properties of water and many compounds of these elements.

The normal covalent bond, not the hydrogen bond, between other atoms and hydrogen differs from all other bonds in that the proton merges almost completely with a cloud on the other atom. This makes our traditional bond formulas of these compounds misleading. Let us compare HF and LiF in several formulations: Lewis dot structures (Figure 1-2a), electron pair cloud (tangent sphere) models (Figure 1-2b), and contour plots of electron cloud densities calculated with molecular orbital theory (Figure 1-2c). The orbital spheres in Figure 1-2b agree with the densities in Figure 1-2c far better than does the dot picture, Figure 1-2a. Ball-and-stick models are also seriously misleading with the hydrogen compounds.

The ability of protons to migrate from cloud to cloud among the highly electronegative atoms is a major feature of acid–base chemistry. In water solution, a Brønsted acid, like HF, may donate protons to water molecules. A Lewis acid, like $Al(H_2O)^{3+}$, can be seen to act like the Brønsted acids. The actual Lewis acid Al^{3+} withdraws electron cloud from the attached water molecules and also repels the H^+, which then jumps to neighboring water molecules. Either action increases the concentration of H_3O^+ and decreases the concentration

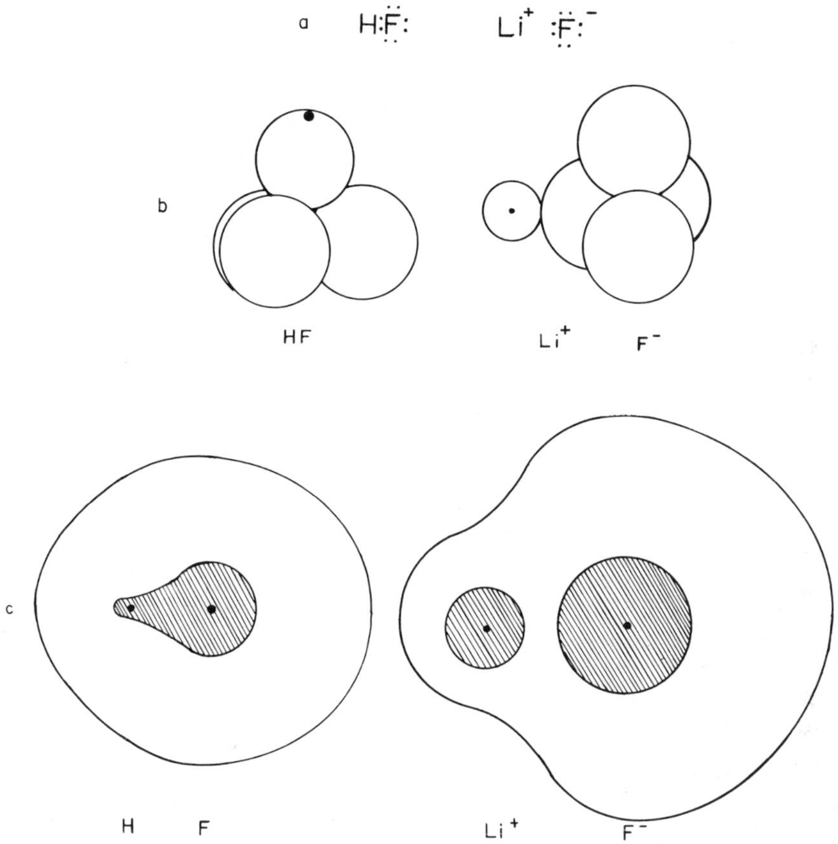

Figure 1-2. (a) Lewis dot structures for LiF and HF. (b) Electron pair cloud (tangent sphere) models of LiF and HF. [See H. A. Bent, *J. Chem. Ed.* **40**, 446 523 (1963).] (c) Contours of electron charge density calculated for LiF and HF. Inner contour is for 0.2 and the outer is for 0.002 atomic unit (6.7 e/Å^3). This shows the ionic nature of LiF, while the nuclei of HF are both within the same inner cloud. [See R. F. Bader, W. H. Henneker, and P. E. Cade, *J. Chem. Phys.* **47**, 3381 (1967); **49**, 1653 (1968); and A. C. Wahl, "Chemistry by Computer," *Sci. Am.* **1970** (April).]

of OH^-. The donor–acceptor nature of the solvent is central to our view of acid–base chemistry in water solutions.

Acid–base concepts share some features with seemingly different types of interactions: complexing, precipitation, and even redox. Comparison of these through theories of Brønsted, Lewis, Usanovich, and Lux–Flood extends our use of the electronic interpretations of

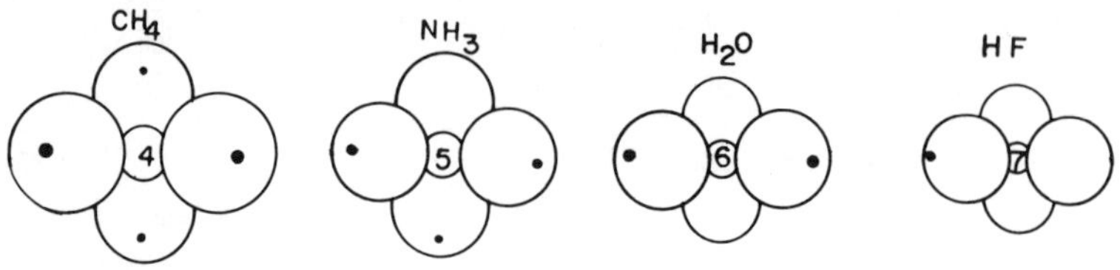

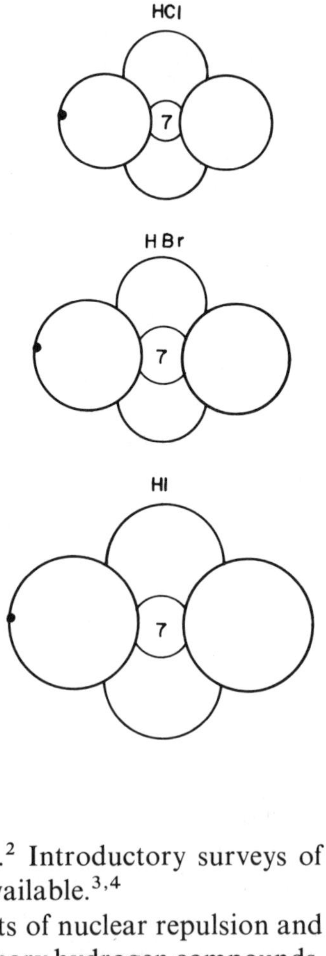

Figure 1-3. Electron pair cloud models of binary hydrogen compounds arranged according to the position of the kernel element in the periodic table. The inner sphere is the kernel with its net positive charge shown. The other spheres are the tetrahedral (sp^3) clouds of spin-paired electrons. The dots represent the protons embedded within electron clouds. These protons are actually too small to put to scale. Features to note: (1) The protons move farther from the nucleus as the kernel charge increases. The molecule becomes more acidic toward water. (2) The cloud size (from covalent radii) becomes smaller as the kernel charge increases, and larger as distance from the nucleus increases while kernel charge is held constant from HF to HI. (3) Two factors affect acidity, kernel charge, from CH_4 to HF, and cloud size, from HF to HI. The compact cloud of HF can bind the proton better than the diffuse cloud of HI.

chemistry. Huheey surveys these theories.[2] Introductory surveys of structural and theoretical principles are available.[3,4]

Figure 1-3 shows the competing effects of nuclear repulsion and electron cloud attraction for a proton in binary hydrogen compounds.

[2]J. E. Huheey, *Inorganic Chemistry*, Harper and Row, New York, 1972, Chapter 6.
[3]G. M. Barrow, *General Chemistry*, Wadsworth Publishing Co., Belmont, California, 1972, Chapters 7 and 11, "Structure" and "Equilibria."
[4]Henry A. Bent, *The Second Law*, Oxford University Press, New York, 1965, Chapter 32, "Thermodynamics of Acid–Base Reactions."

2. The Oxy-Acid–Bases

The same arguments concerning size and charge effects in the binary hydrogen compounds of Figure 1-3 apply to the major class of acid–bases, the oxy-acid–bases. In these, a protonated oxygen is attached to other atoms, usually electron-withdrawing atoms or groups. The more withdrawing, the more acidic the proton, the more favorable is its transfer to the oxygen cloud on a water molecule, for example, compare sulfate(IV) and sulfate(VI) acids:

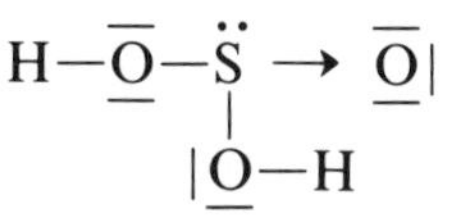

weak acid

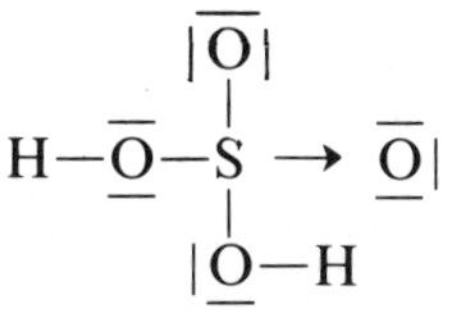

strong acid

H_2SO_4 has one more coordinated oxide than H_2SO_3. This withdraws more electron cloud from the S, and through it, from the OH groups. Another way to approach this is to note that the oxidation state of S is six in H_2SO_4, so that the more exposed positive kernel of the S withdraws cloud more strongly from OH groups than does S(IV).

Similar reasoning can be applied to other series, for example:

Acid	HOCl	HOClO	$HOClO_2$	$HOClO_3$
pK_a	7.5	2.0	strong	strong

Carboxylic acids have the withdrawing coordinated oxide also,

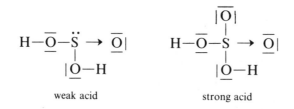

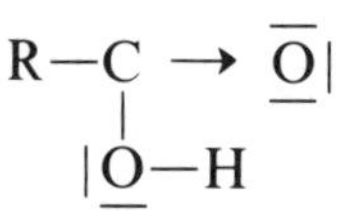

The pK_a is 4–5 when R is aliphatic. But if other withdrawing groups are added in R, acidity increases:

Acid	acetic	chloroacetic	dichloroacetic	trichloroacetic
pK_a	4.76	2.86	1.3	0.7

Thus we see that for most acids in water, the proton shifts between electron clouds on oxygen in water and in other molecules with slightly different environments.

3. Brønsted Acid–Base Pairs versus "Hydrolysis"

Metal ions in solution are truly oxy-acids. Many of these have large kernels and use d^2sp^3 orbitals for six-coordinate bonding. They can hold H_2O as well as OH^- or O^{2-} of the smaller oxy-acids. Ions like Fe^{3+} strongly hold six water molecules and withdraw electron cloud from the oxygen. This produces acid behavior in no way different from that of the other acids we have considered,

$$Fe(OH_2)_6^{3+} + H_2O \rightleftharpoons Fe(H_2O)_5OH^{2+} + H_3O^+$$

$$K_a = 7 \times 10^{-3}$$

Thus, the ferric ion is considerably more acidic than acetic acid. Dilute ferric solutions are often yellow due to the hydroxopentaquo ion. Addition of HNO_3 or $HClO_4$ shifts the equilibrium to the left, giving the pale violet hexaquo ion.

Note that we do not ever need to refer to "hydrolysis" to explain acid–base behavior in water. It only confuses the issue. All the Brønsted acids and bases follow identical patterns of proton transfer. There is no chemical reason to call some of these examples dissociation and some hydrolysis:

Acids:

$$HX + H_2O \rightleftharpoons H_3O^+ + X^-$$
$$HF + H_2O \rightleftharpoons H_3O^+ + F^-$$
$$NH_4^+ + H_2O \rightleftharpoons H_3O^+ + NH_3$$
$$HSO_4^- + H_2O \rightleftharpoons H_3O^+ + SO_4^{2-}$$
$$Fe(H_2O)_6^{3+} + H_2O \rightleftharpoons H_3O^+ + Fe(H_2O)_5OH^{2+}$$

Bases:

$$B + H_2O \rightleftharpoons OH^- + HB$$
$$NH_3 + H_2O \rightleftharpoons OH^- + NH_4^+$$
$$CO_3^{2-} + H_2O \rightleftharpoons OH^- + HCO_3^-$$
$$HCO_3^- + H_2O \rightleftharpoons OH^- + H_2CO_3$$

Each acid has a conjugate base lacking one proton, and each base has a conjugate acid having one added proton. Thus, we speak of an acid–base pair, since they must exist together at equilibrium.

The relative strengths of acid–base pairs have been estimated and are shown diagrammatically in Figure 1-4. An acid can protonate, to an equilibrium concentration, the conjugate base of any acid below it. Thus, the strong acids protonate water to give H_3O^+, and the strong bases are protonated by water to form OH^-. Further discussions of these large topics can be found in general texts and in the references cited in footnotes 1–4 as well as in the figure legends.

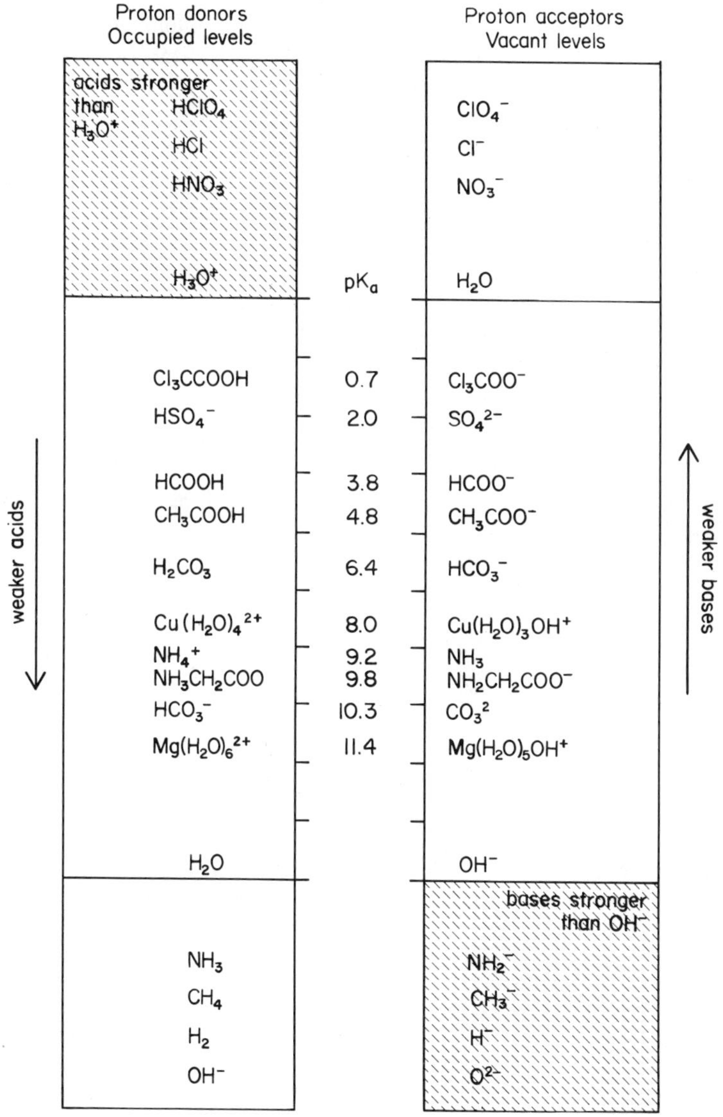

Figure 1-4. Aqueous proton levels shown as free energy levels. [See Henry A. Bent, *The Second Law*, Oxford University Press, New York, 1965, Chapter 32; and R. W. Gurney, *Ionic Processes in Solution*, McGraw-Hill, New York, 1953 (reprinted Dover, 1962), p. 133.]

Problems[5]

1. Classify the following substances as to their behavior in dilute water solutions as strong acid or base, weak acid or base, practically neutral. Use Figure 1-4 and analogies based on the periodic table:

   HBr, KOH, KBr, KNH_2, C_4H_7OOH (butyric acid), NH_4NO_3, $Ca(C_2H_3O_2)_2$,

   $CrCl_3{\cdot}6H_2O$

2. Write chemical equations for the net ionic reactions occurring when:
   a. Trisodium phosphate in water produces a very basic solution.
   b. Dilute solutions of ammonium chloride and sodium hydroxide are mixed.
   c. Ammonia acts weakly acidic when passed into saturated $NaOH_{aq}$.
   d. Acetic acid can be a base (proton acceptor) in pure liquid H_2SO_4.
   e. Strong acids form $H_3O^+_{aq}$ in water, and NH_4^+ in liquid NH_3.
   f. Sodium reacts with either water or methanol to give H_2 gas. What might be the self-ionization reaction of methanol?

---

3. Each of these groups is arranged from the strongest to the weakest acid in water solutions. Draw Lewis dot structures and explain these trends on the bases of size and charge effects.
   a. H_2Te, H_2Se, H_2S, H_2O. b. H_2CO_3, H_2SiO_3.
   c. HOCl, HOBr, HOI. d. HNO_3, HNO_2, H_2CO_3, HBO_2.

4. When solutions of Fe(III) compounds are mixed with solutions of Na_2CO_3, $NaHCO_3$, or $NaC_2H_3O_2$, similar precipitates of hydrated ferric oxide ["$Fe(OH)_3$"] form. Explain.

5. Each of the following groups is arranged from strong to weak acid. Explain the opposite trends. H_2Te, H_2Se, H_2S; and H_2SO_4, H_2SeO_4, H_2TeO_4.

6. When solutions of chromium(III) nitrate and potassium bicarbonate are mixed, a gas bubbles off and a light-green precipitate forms. Write a net ionic equation and explain the acid–base reactions.

[5] The problems after most chapters will be divided into two or three groups: (a) up to the single line: drill on basic ideas and operations; (b) from the single to the double line: more advanced or more complex situations; (c) after the double line: applications to special fields, not necessarily more difficult than sets (a) or (b).

2 Chemical Equilibrium Relations and the pH System

1. The Equilibrium Constant

Chemists developed sufficiently sensitive methods of analysis in the 19th century to discover that many reactions "stop" somewhere short of completion. Berthelot and St. Gilles in 1862 showed that esterification reaches such a limit:

$$\text{ethanol} + \text{acetic acid} \rightleftharpoons \text{ethyl acetate} + \text{water}$$

Their experimental yields for various mixtures were later shown to give a roughly constant quotient of the form

$$Q = \frac{[\text{ethyl acetate}][\text{water}]}{[\text{ethanol}][\text{acetic acid}]} = 4$$

From a modern reinvestigation of this equilibrium without strong acid catalysts present,[1] the two experiments summarized in Table 2-1 demonstrate how different mixtures achieve this quotient after reacting until an equilibrium state is reached. The figures are given on the basis of 100 moles of materials (mole percentage) at 25°C. Note that even though the proportions differ, the quotient at equilibrium is about the same. (The medium here is not a dilute water solution, and the quotients found did differ with the medium, varying from 1.9 to 4.6; see reference cited in footnote 1.)

[1] A. Darlington and W. B. Guenther, *J. Chem. Eng. Data* **12**, 605 (1967).

Table 2-1

	HA	EtOH $\rightleftharpoons$	EtA	H_2O	Q	Moles of each shifted
Experiment 1						
Moles taken	0	8.68	2.58	88.7		1.8←
Moles at equilibrium	1.80	10.47	0.785	86.9	3.62	
Experiment 2						
Moles taken	9.43	18.6	41.4	30.6		3.4←
Moles at equilibrium	12.9	22.0	37.9	27.2	3.63	

An example with simpler numbers will clarify these equilibrium shifts. Table 2-2 is for a hypothetical case of Q about 9. No matter what proportions are taken to start, the changes linked by the reactions proceed until the Q value is attained. This is a statement of the "law of mass action," which was formulated by Guldeberg and Waage in 1865. Van't Hoff in 1877 gave the kinetic explanation that such a relation must follow if rates of opposing reactions become equal when the equilibrium concentrations are reached.

The mathematical form of the mass action law for a general reversible chemical reaction

$$aA + bB + \cdots \rightleftharpoons xX + yY + \cdots$$

is

$$K^{\circ}_{\text{eq}} = \frac{(X)^x(Y)^y \cdots}{(A)^a(B)^b \cdots} \tag{2-1}$$

Table 2-2

	A +	B $\rightleftharpoons$	C +	D	Q	Moles of each shifted
1. Start	0	0	4	4		
At equilibrium	1	1	3	3	9.0	1←
2. Start	4	4	0	0		
At equilibrium	1	1	3	3	9.0	3→
3. Start	1	1	2	6		
At equilibrium	1.1	1.1	1.9	5.9	9.2	0.1←

Strictly constant values for this quotient are obtained only at constant temperature and pressure and by use of the thermodynamic activity for each term in the mass action expression (2-1). This is related to molarity in the solution by

$$\text{activity of solute species A} = (\text{A}) = f[\text{A}] \tag{2-2}$$

Parentheses indicate activity and brackets indicate molarity. The activity coefficient f depends upon the nature of the medium as well as upon A. For ionic species of small radius (2–5 Å), the Debye–Hückel expression for f is valid in dilute solutions:

$$\log f = \frac{-0.509z^2\sqrt{I}}{1 + \sqrt{I}} \tag{2-3}$$

Here, z is the charge on the ion and I is the ionic strength of the medium, defined as half the sum of cz^2 for *all* the ions present:

$$I = \frac{1}{2}\sum_i c_i z_i^2 \tag{2-4}$$

Here c_i is the concentration of ion i. Un-ionized molecules are assumed to have $f = 1$, and to be rather unaffected by ions around them. Equation (2-3) is fairly useful up to ionic strength about 0.1 *M*. As I approaches zero, f approaches 1 and the activity approaches the molarity. Only when I is less than about 0.001 *M* can molarities be used for activities without errors greater than several percent. In the calculations of this book, we shall follow the practice of obtaining a corrected, conditional, equilibrium constant for the ionic strength required. Then molarities can be used (but only for that one ionic strength). The conditional constant K is obtained from published activity constants $K°$ as follows. From Eqs. (2-1) and (2-2),

$$K^\circ_{\text{eq}} = \frac{(f_x[\text{X}])^x(f_y[\text{Y}])^y\cdots}{(f_a[\text{A}])^a(f_b[\text{B}])^b\cdots} \tag{2-5}$$

At constant ionic strength, the f values are constant. We collect all the constant terms to form the conditional constant:

$$K_{\text{eq}} = K^\circ_{\text{eq}}\frac{f_a^a f_b^b\cdots}{f_x^x f_y^y\cdots} = \frac{[\text{X}]^x[\text{Y}]^y\cdots}{[\text{A}]^a[\text{B}]^b\cdots} \tag{2-6}$$

The obtaining and use of numerical values for these K's will be demonstrated in example problems to follow as the need arises.

Activity coefficients are found in Appendix A-1, Tables A-1 and A-2, and in Kielland[2] and Hamer.[3]

2. Water

For a complete description of acid–base equilibria in water, one must include the acid–base behavior of the solvent itself:

$$H_2O + H_2O \rightleftharpoons H_3O^+ + OH^- \quad (2\text{-}7)$$

Protons can move from the cloud on one oxygen to that on a neighboring oxygen. The acid species formed is highly hydrated beyond the H_3O^+ to form some $H_5O_2^+$, $H_7O_3^+$, and $H_9O_4^+$, the last being the most abundant. A more general equation is then

$$H_2O + nH_2O \rightleftharpoons H^+(H_2O)_n + OH^-$$

At a given moment, about two billionths of the water molecules are in the ionic form on the right, that is, 10^{-7} mol out of the 55.5 mol in a liter of water. The equilibrium constant for this self-ionization is

$$K^\circ_{eq} = \frac{(H^+(H_2O)_n)(OH^-)}{(H_2O)^{n+1}} \quad (2\text{-}8)$$

We take the activity of pure water as 1 and obtain the conventional ion product constant for the self-ionization of water,

$$\begin{aligned} K^\circ_w &= (H^+_{aq})(OH^-) = f_+ f_- [H^+_{aq}][OH^-] \\ K_w &= [H^+_{aq}][OH^-] = K^\circ_w / f_+ f_- = \mathbf{H \cdot OH} \end{aligned} \quad (2.9)$$

Here we have written $[H^+_{aq}]$ for the sum of all forms of hydrated protons. We shall now adopt the symbol **H** for this molarity and **OH** for the term $[OH^-]$.

In pure water, the hydronium and hydroxide ions are effectively at infinite dilution and we take their f values as 1. The true K°_w can be found as a function of temperature:

Temperature, °C	0°	15°	25°	40°	60°
$K^\circ_w \times 10^{14}$	0.114	0.450	1.008	2.92	9.61

---

[2] J. Kielland, *J. Am. Chem. Soc.* **59**, 1675 (1937).

[3] W. J. Hamer, *Theoretical Mean Activity Coefficients of Strong Electrolytes in Aqueous Solutions from 0 to 100° C*, NBS No. 24, U. S. Government Printing Office, Washington, D. C., 1968.

Thus, we see that K_w° varies nearly 100-fold from 0 to 60°C. Note that it is pure chance that at 25°C, near room temperature, we have $K_w^\circ \cong 1.0 \times 10^{-14}$. The negative log, pK_w°, is equal to 13.996, or 14.00. There is no magic in the rounded value 7.00 for the pH of ultrapure water at 25°C. Actually, it is 6.998. The pH of pure water is 7.47 at 0° and 6.51 at 60°C. Remember that the definition of neutrality in water is not that the pH is seven, but that the hydronium concentration is equal to the hydroxide concentration.

When "inert" salt ions like K^+Cl^- are placed in water, the activities of the H_3O^+ and OH^- are affected. If we take as fact that the K_w° must remain constant, then a change in f values means that molarities must change in such a way as to keep the activity product constant. This has been found experimentally. The molarity product K_w does vary at 25° as KCl is added to water:

M KCl	0.05	0.10	0.25	0.50	1.00	2.00
$K_w \times 10^{14}$	1.5	1.6	1.9	2.1	1.7	1.4

Thus, we see that the ionic atmosphere affects the degree of ion formation of water.

These data on the temperature and ionic strength effects on K_w show up the futility of doing calculations with more than one significant digit (or to $\pm 10\%$) when T and ionic strength are unknown.

3. Monoprotic Acids

When a protonic acid HX is placed in water, it reacts partially (or completely if the acid is strong) to form some ions. The degree of hydronium hydration will not affect the calculations when water activity 1 is assumed for dilute solutions.

$$HX + H_2O \rightleftharpoons H_3O^+ + X^-$$

$$K_{eq}^\circ = \frac{(H_3O^+)(X^-)}{(H_2O)(HX)} = \frac{[H_3O^+][X^-]f_+f_-}{[HX]} = K_a f_+ f_- \qquad (2\text{-}10)$$

Here again, we take activity *unity* for water (for dilute solutions) and activity coefficient *unity* for unchanged species like HX. The K_a is the conditional molarity acidity constant. This varies with temperature and ionic strength. For example, the data for acetic acid are

Temperature, °C	0°	15°	25°	40°	60°
$K_a^\circ \times 10^5$	1.66	1.74	1.75	1.70	1.54

With added NaCl, the molarity K_a was found for acetic acid as follows:

M NaCl	0	0.06	0.1	0.2	0.5	1	2	3
$K_a \times 10^5$	1.75	2.6	2.8	3.1	3.3	3.2	2.5	1.8

4. Bases and the K_b

When sources of a Brønsted base (see Chapter 1) are added to water, their equilibria can be expressed by the traditional K_b using (H_2O) as unity as before, e.g.,

$$NH_3 + H_2O \rightleftharpoons NH_4^+ + OH^-$$

$$K_b^\circ = \frac{(NH_4^+)(OH^-)}{(NH_3)}$$

However, modern usage is to list only acidity constants in tables, or their inverse, the formation constants (base + proton).[4] In this case, we can show why this is enough by examining the acidity constant expression for the conjugate acid NH_4^+,

$$NH_4^+ + H_2O \rightleftharpoons NH_3 + H_{aq}^+$$

$$K_a^\circ = \frac{(H_{aq}^+)(NH_3)}{(NH_4^+)}$$

If we multiply the K_w° expression by $1/K_b^\circ$, we get K_a°:

$$\frac{K_w^\circ}{K_b^\circ} = \frac{(H_{aq}^+)(OH^-)}{(NH_4^+)(OH^-)/(NH_3)} = \frac{(H_{aq}^+)(NH_3)}{(NH_4^+)} = K_a^\circ$$

Thus, either K_a° or K_b° gives us all the equilibrium information about an acid–base conjugate pair,

$$K_a^\circ K_b^\circ = K_w^\circ$$

[4]See L. G. Sillen and A. E. Martell, *Stability Constants of Metal Ion Complexes*, Special Publications No. 17, 1964, and No. 25 (Supplement), 1971, The Chemical Society, London.

The K_b° for ammonia is $10^{-4.76}$. Therefore the K_a° for NH_4^+ must be $10^{-9.24}$. Their product is $10^{-14.00}$.

5. The pH System

Using the equations developed so far, we can examine the interrelations of the species present at equilibrium in pure water and in solutions of acids and bases. Clearly equation (2-9) requires that the product of the two variables **H** and **OH** be constant. As one rises, the other must fall. For the moment, consider roughly, using molarities for activities, the possible ranges of **H** and **OH**. Neither can be zero and satisfy equation (2-9). At the other extreme, one liter can contain a limited amount of solute. One hundred moles, or 4000 g of NaOH will not fit in a liter of solution. Let us take 10 mol/liter as about the maximum possible. Now we look at the total range between 10 *M* strong acid and 10 *M* strong base. In 10 *M* HCl solution we take **H** as approximately 10 *M*. (This hardly qualifies as a dilute aqueous solution.) What is **OH**? Equation (2-9) requires

$$\mathbf{OH} = K_w/\mathbf{H} = 10^{-14}/10 = 10^{-15}$$

and in 10 *M* NaOH solution,

$$\mathbf{H} = K_w/\mathbf{OH} = 10^{-14}/10 = 10^{-15}$$

This illustrates the reciprocality of these variables. In more dilute solution, we are on firmer ground, and can derive this set of **H** and **OH** values for strong acid and base solutions in Table 2-3.

The **pH** system, proposed by Sorensen in 1909, is seen from Table 2-3 to be the use of the negative logarithm of **H** to express the wide range of acidities over about 17 powers of ten. Early workers assumed that they were dealing with the molarity of H_3O^+. Equation (2-8) and the data on inert salt effects shows that this is too simple. Potentiometric devices like the pH meter respond to the acidity of water solutions in a way that closely approximates what we believe to be the activity, (H_{aq}^+). We are forced into this vague statement by the fact that there is really no rigorous way to measure the activity of one kind of ion alone. What all physical measurements give us is the mean activity of the electrolyte in solution, H^+, Cl^- or Na^+, OH^-. For

Table 2-3

M	H	OH	pH
Strong acid HX			
10	10	10^{-15}	−1
1	1	10^{-14}	0
0.1	10^{-1}	10^{-13}	1
0.02	2×10^{-2}	5×10^{-13}	1.7
0.001	10^{-3}	10^{-11}	3
Strong base MOH			
10	10^{-15}	10	15
1	10^{-14}	1	14
0.1	10^{-13}	10^{-1}	13
0.02	5×10^{-13}	2×10^{-2}	12.3
0.001	10^{-11}	10^{-3}	11

our purposes we shall treat the pH meter readings as $-\log(\mathrm{H}^+_{\mathrm{aq}})$, where

$$\text{hydronium activity} = (\mathrm{H}^+_{\mathrm{aq}}) = \mathbf{H}f_+$$

and the f_+ value will have to be the best we can find by the approximate methods, equation (2-3), or from tables (see references in footnotes 2 and 3). In solutions of ionic strength below about 0.1 *M*, this approach will give us agreement on the order of 0.01 pH unit between measured and calculated values from the K_a° expressions. The problems of definition and choice of standards are presented in detail by Bates.[5] To avoid most of the problems, a standard pH scale is *defined* in terms of pure standard solutions (Table 2-4). Thus, a measured pH relates to these standards and does not depend upon any supposed theory about what it *is*.

We now consider examples to illustrate the meaning and uses of the ideas introduced in this chapter.

Example 1. What are **H**, the molarity **pH**, and the pH likely to be read from a properly standardized pH meter for 0.100 *M* HCl?

By definition, strong acids are assumed to be 100% in ionic form in dilute solutions. Thus, we have **H** at 0.100 *M*, and the approximate **pH** is 1.0. Next, we calculate the hydronium activity coefficient

[5] Roger Bates, *The Determination of pH*, 2nd ed., Wiley–Interscience, New York, 1973.

Table 2-4. National Bureau of Standards Primary Standards for pH with Concentrations Expressed in Molalities (mol/kg)[a,b]

Temperature, °C	KH tartrate (satd. at 25°C)	KH_2 citrate (m = 0.05)	KH phthalate (m = 0.05)	KH_2PO_4 (m = 0.025), Na_2HPO_4 (m = 0.025)	KH_2PO_4 (m = 0.008695), Na_2HPO_4 (m = 0.03043)	Borax (m = 0.01)	$NaHCO_3$ (m = 0.025), Na_2CO_3 (m = 0.025)
0	—	3.863	4.003	6.984	7.534	9.464	10.317
5	—	3.840	3.999	6.951	7.500	9.395	10.245
10	—	3.820	3.998	6.923	7.472	9.332	10.179
15	—	3.802	3.999	6.900	7.448	9.276	10.118
20	—	3.788	4.002	6.881	7.429	9.225	10.062
25	3.557	3.776	4.008	6.865	7.413	9.180	10.012
30	3.552	3.766	4.015	6.853	7.400	9.139	9.966
35	3.549	3.759	4.024	6.844	7.389	9.102	9.925
38	3.548	3.755	4.030	6.840	7.384	9.081	9.903
40	3.547	3.753	4.035	6.838	7.380	9.068	9.889
45	3.547	3.750	4.047	6.834	7.373	9.038	9.856
50	3.549	3.749	4.060	6.833	7.367	9.011	9.828
55	3.554	—	4.075	6.834	—	8.985	—
60	3.560	—	4.091	6.836	—	8.962	—
70	3.580	—	4.126	6.845	—	8.921	—
80	3.609	—	4.164	6.859	—	8.885	—
90	3.650	—	4.205	6.877	—	8.850	—
95	3.674	—	4.227	6.886	—	8.833	—

[a] These pH values apply only when the solutions are properly prepared from purified materials and protected from the CO_2 of the air.

[b] R. G. Bates, *The Determination of pH*, 2nd ed., Wiley–Interscience, New York, 1973; *J. Res. Nat. Bur. Std.* **66A**, 179 (1962); B. R. Staples and R. G. Bates, *J. Res. Nat. Bur. Std.* **73A**, 37 (1969).

from equation (2-3) or get it from the Kielland table in Appendix A-1 as 0.83. The estimated hydronium activity is

$$(H^+_{aq}) = \mathbf{H}f_+ = 0.100(0.83) = 0.083$$

This gives the activity pH as $-\log(0.083)$ or 1.08. The NBS value is 1.088 for use of this solution as a secondary standard (p. 71 in Bates, see footnote 5).

Example 2. Find the **H** and the pH meter reading expected for a solution which is 0.100 *M* in acetic acid and 0.100 *M* in sodium acetate.

All sodium compounds are taken to be strong electrolytes, so that we have 0.100 *M* Na^+ and 0.100 *M* acetate A^- from this source. Much smaller concentrations of ions come from the weak electrolytes water and acetic acid. So, try a first approximation of 0.100 for the ionic strength [equation (2-4)]. Get the activity coefficients and put them into equation (2-10) with our approximate values of 0.100 for acetate and acetic acid, and find **H**:

$$K_a^\circ = 1.76 \times 10^{-5} = \mathbf{H}\frac{(0.100)}{(0.100)}(0.83)(0.76)$$

$$\mathbf{H} = 2.8 \times 10^{-5}$$

$$(H^+_{aq}) = 2.8 \times 10^{-5}(0.83) = 2.3 \times 10^{-5}$$

$$\text{pH} = 5 - \log 2.3 = 4.64$$

This agrees with experimental measurements (p. 100 in Bates, see footnote 5). It also justifies our approximations since well under 1 % of the acetic acid has formed ions.

To clarify what we have done here, we make an inventory of what is present in Table 2-5. The small values of **H** and **OH** *relative* to the

Table 2-5

Species	Put into the solution	Present at equilibrium
Na^+	0.100 *M*	0.100 *M*
$C_2H_3O_2^-$	0.100 *M*	0.100 + **H** = 0.100 *M*
$HC_2H_3O_2$	0.100 *M*	0.100 − **H** = 0.100 *M*
H_2O	55.4 *M*	55.4 − **OH** = 55.4 *M*
H	—	2.8×10^{-5} *M*
OH	—	$K_w/\mathbf{H} = 3.6 \times 10^{-10}$ *M*

larger concentrations are what make our approximations good ones. This is by no means always possible to assume correctly. A complete and systematic treatment of these equilibrium calculations is presented in the chapters to follow. Here we reiterate that, although pH lacks a rigorous meaning, it is a practical tool for expressing the relative acidities of solutions under clearly stated conditions. The interpretation of pH measurements is required for intelligent use of laboratory and field data on solutions. The importance of these equilibria in controlling processes in life chemistry, industrial chemistry, and geochemistry makes this study vital to all workers in the sciences.

Problems

1. Write equilibrium equations and the constant expressions for any acid–base equilibria which must become satisfied in dilute water solutions of the following compounds:
   a. Formic acid, HCOOH.
   b. Sodium formate, Na^+, $HCOO^-$.
   c. Trimethylamine.
   d. H_2SO_4, one H is strong, the second is weak.

2. What are the concentrations of major species present in the following water solutions, and what is the pH assuming activity coefficients of 1?
   a. 0.0020 *M* HNO_3.
   b. 0.008 *M* KOH.
   c. 0.015 *M* HI.
   d. 0.003 *M* $NaNH_2$.

---

3. Calculate the **H**, molarity, before and after dissolving 0.05 mole of NaCl in one liter of 0.20 *M* acetic acid at 25°.

4. What is the pH of ultrapure water at 15 and at 40°?

5. Calculate the molarity **H** and the activity pH of 0.050 and 0.10 *M* KCl solutions at 25°C.

6. Find the molarity **H** in a water solution which produces a reading of 2.30 at the pH meter and has ionic strength giving f_+ of 0.80.

7. Take the K_{eq} as 4.00 and calculate the equilibrium number of moles of each substance after mixing 1.00 mol of each of the four compounds involved in the ethyl acetate esterification. Compare the yield when only 1 mol each of ethanol and acetic acid are mixed.

3 Monoprotic Acid–Base Equilibria in Water Solutions

The various species with acid–base properties can be related to each other by the equations expressing their equilibria and by equations expressing conservation of matter: material, charge, and proton balancing equations. When initial (analytical) concentrations of solutes are given, these equations serve to determine as many unknowns as we have equations. Molarities, not activities, must be used in these balances, so we require the conditional (molarity) equilibrium constants appropriate to the solution under consideration.

1. A Complete Equation Treatment

Let us reexamine the simplest case, strong acid solution, Example 1 in Chapter 2. In this 0.100 *M* HCl solution, it is not exact to set **H** equal to the molarity of HCl. The **OH** cannot be zero, therefore water must produce some **H** and **OH**. Call this amount $y\,M$. Now the complete situation is expressed by

$$\mathbf{H} = 0.100 + y, \qquad \mathbf{OH} = y$$

Using the K_w from Chapter 2 for 0.1 ionic strength and 25°, we obtain

$$K_w = 1.6 \times 10^{-14} = \mathbf{H}{\cdot}\mathbf{OH} = (0.100 + y)(y)$$

$$y = \mathbf{OH} = 1.6 \times 10^{-13}, \qquad \mathbf{H} = 0.100$$

While this treatment has not changed our approximate result found before, this is not always the case. Consider a very dilute solution, 1.00×10^{-7} M HCl. Now, using the zero-ionic-strength K_w° values we get

$$K_w^\circ = 1.008 \times 10^{-14} = (1.00 \times 10^{-7} + y)(y)$$

Solving with the quadratic formula and taking only the positive root gives

$$y = 0.62 \times 10^{-7} = \mathbf{OH}$$

$$\mathbf{H} = 1.62 \times 10^{-7}$$

This time, the approximation that all the **H** comes from the added HCl gives the silly answer that the solution is neutral. Note that the HCl does not simply add to the 10^{-7} M **H** from water. The fallacy of the approximate method is also shown by application of the electroneutrality principle, charge balancing, which we shall use a great deal. It states the rather obvious condition that (in macro quantities of solution) positive and negative ions must balance each other. This means we count Ca^{2+} as 2 mol of plus charge. In the present example, we set all the possible positive charges equal to all possible negative ions. There are only three possible ions in our solution: **H**, **OH**, and $[Cl^-]$:

$$\mathbf{H} = [Cl^-] + \mathbf{OH}$$

$$1.00 \times 10^{-7} = 1.00 \times 10^{-7} + 1.00 \times 10^{-7} \text{ (approximate)}$$

$$1.62 \times 10^{-7} = 1.00 \times 10^{-7} + 0.62 \times 10^{-7} \text{ (complete)}$$

Weak Monoprotic Acid–Bases. It has been traditional in elementary texts to use molarity equilibrium constant expressions and to make approximations for the numerical values of the concentrations of all species except the one to be calculated, usually **H**. This is a valid approach, but difficult for students because of the variety of assumptions used in making the approximations. Let us start by looking at the complete treatment and then showing under what conditions certain approximations will be justified. Until experience is achieved, the best method is to try the approximations and then check the result in the complete equation.

Here we list terms and symbols to be used:

C_a	the analytical concentration of acid HX put into the solution
C_b	the analytical concentration of the conjugate base put into the solution
HX	the concentration of HX actually present at equilibrium
X	the concentration of X^- actually present at equilibrium
H	the concentration of hydrated protons at equilibrium
OH	the concentration of OH^- at equilibrium
D	**H** − **OH**, the net acidity, or basicity (when negative), at equilibrium

To illustrate these, in 0.100 *M* acetic acid we have $C_a = 0.100\,M$, $C_b = 0$, **HX** is slightly less than 0.100 *M*, and the **H**, **OH**, and **X** are small unknown concentrations dependent upon the value of K_a. If we add sodium acetate to make its total in solution 0.100 *M*, then C_b becomes 0.100 *M* and the equilibrium values of all the other components are shifted to new concentrations to satisfy K_a. Now we derive the general expression relating these quantities.

For solutions which may have any values of C_a and C_b including zero, we can write the material balance, which simply states that the total of X at equilibrium must equal the total put into the solution,

$$C_a + C_b = \mathbf{HX} + \mathbf{X}$$

We also write a charge balance taking into account that every X^- from the C_b added must also be balanced by M^+ (Na^+, for example, if sodium acetate is used):

$$\mathbf{H} + C_b = \mathbf{OH} + \mathbf{X}, \qquad \mathbf{X} = C_b + \mathbf{D}$$

Eliminate C_b between the charge and material balances to get

$$\mathbf{HX} = C_a - \mathbf{D}$$

Thus, we have obtained expressions for the equilibrium **HX** and **X** in terms of the analytical concentrations which we often know. Putting these into the equilibrium constant expression solved for **H**, equation (2-10), gives the general relation we need:

$$\mathbf{H} = K_a\mathbf{HX}/\mathbf{X} = K_a(C_a - \mathbf{D})/(C_b + \mathbf{D}) \tag{3-1}$$

We can use the K_w relation, equation (2-9), to get this in terms of the single unknown **H**:

$$\mathbf{H} = K_a(C_a - \mathbf{H} + \mathbf{OH})/(C_b + \mathbf{H} - \mathbf{OH})$$

$$\mathbf{H} = K_a(C_a - \mathbf{H} + K_w/\mathbf{H})/(C_b + \mathbf{H} - K_w/\mathbf{H}) \qquad (3\text{-}2)$$

This is a cubic equation in **H**, and will be useful in this form for the purposes at hand.[1]

This expression reduces to the common approximations as follows. Take first the case of acid solutions alone, $C_b = 0$. Assume that **H** and **OH** are negligible with respect to C_a in the numerator of equation (3-2), and that **H** is much larger than **OH** in the denominator. This gives us

$$\mathbf{H} = K_a C_a/\mathbf{H}, \qquad \mathbf{H} = (K_a C_a)^{1/2} \qquad (3\text{-}3)$$

For base solution alone, $C_a = 0$, and **H** smaller than **OH**, we get

$$\mathbf{H} = K_a \mathbf{OH}/C_b, \qquad \mathbf{H} = (K_a K_w/C_b)^{1/2} \qquad (3\text{-}4)$$

For buffers, C_a and C_b both nonzero, similarly, we get

$$\mathbf{H} = K_a C_a/C_b \qquad (3\text{-}5)$$

Comparing these results with the full equation (3-2), we can deduce the conditions under which (3-3)–(3-5) will be invalid: (a) when either **H** or **OH** is large—rather strong acids or bases; and (b) when C_a and/or C_b are small—very dilute solutions.

2. Graphical Illustrations

These effects are shown graphically in several ways in Figures 3-1 to 3-4. Figure 3-1 shows the effect of acid strength with K_a values ranging from 1 to 10^{-14}. The **pH** was calculated (in molarity units) from the full equation (3-2) for 0.100 *M* solutions of acid, bases, and buffers (0.100 *M* each of the conjugate acid and base). The approximate equations apply in the central, straight-line regions, as is seen by taking

[1]To show the complete generality of equation (3-2), we can even get the equations for strong acid or base solutions: For strong acid, let K_a grow very large and rearrange to get

$$(\mathbf{H}/K_a)(C_b + \mathbf{H} - \mathbf{OH}) \rightarrow 0 = C_a - \mathbf{H} + \mathbf{OH}, \quad \text{and} \quad \mathbf{H} = C_a + \mathbf{OH}$$

For strong bases, let $K_a \rightarrow 0$, giving

$$(K_a/\mathbf{H})(C_a + \mathbf{OH} - \mathbf{H}) = 0 = C_b + \mathbf{H} - \mathbf{OH}, \qquad \mathbf{OH} = C_b + \mathbf{H}$$

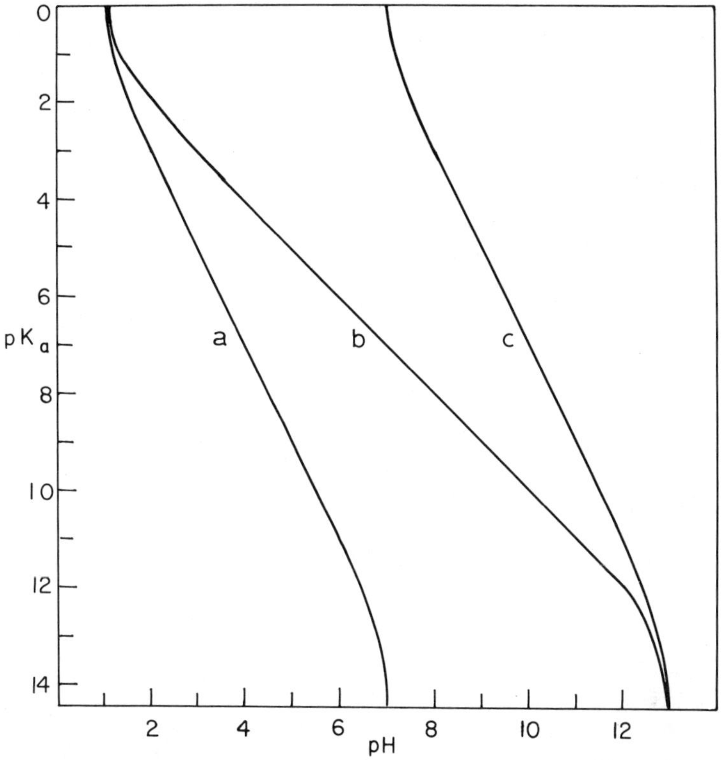

Figure 3-1. The variation of **pH** with acid strength pK_a for 0.10 M solutions of (a) pure weak acid HX alone; (b) buffers, conjugate acid and base each 0.10 M; (c) pure conjugate base alone.

logs: From equation (3-3), $-\log \mathbf{H} = -\frac{1}{2}\log K_a - \frac{1}{2}\log C_a$, or, with $p = -\log$, and for 0.100 M ($\log C_a = -1$),

$$\mathbf{pH} = \tfrac{1}{2}\mathrm{p}K_a + 0.5 \tag{3-6}$$

From equation (3-4),

$$\mathbf{pH} = \tfrac{1}{2}\mathrm{p}K_a + 7 - 0.5$$

From equation (3-5),

$$\mathbf{pH} = \mathrm{p}K_a$$

The plots of pK_a versus **pH** have the slopes shown in Figure 3-1 between pK_a values about 3 and 11. With the stronger acids and bases the other terms in equation (3-2) are required and produce the curvatures.

Let us contrast the actual calculations for the two regions of Figure 3-1.

Example 1. For pure 0.100 M acetic acid solution, we use K_a° in equation (3-3),

$$\mathbf{H} = (K_a C_a)^{1/2} = [1.75 \times 10^{-5}(0.100)]^{1/2} = 1.34 \times 10^{-3}$$

$$\mathrm{p}\mathbf{H} = 2.87$$

For pure conjugate base, 0.100 M sodium acetate, we now use the conditional constants K_a and K_w for 0.1 M ionic strength in equation (3-4),

$$\mathbf{H} = (K_a K_w / C_b)^{1/2} = [2.8 \times 10^{-5}(1.6 \times 10^{-14})/0.100]^{1/2}$$

$$= 2.12 \times 10^{-9}$$

$$\mathrm{p}\mathbf{H} = 8.67$$

In neither case will use of the full equation (3-2) give a different result to two significant digits.

For the buffer, $C_a = C_b = 0.100$ in equation (3-5) gives

$$\mathbf{H} = K_a = 2.8 \times 10^{-5} \qquad \text{and} \qquad \mathrm{p}\mathbf{H} = 4.55$$

Example 2. Now consider 0.100 M trichloroacetic acid, $K_a = 0.2$. The approximate equation (3-3) gives $\mathbf{H} = 0.14\ M$. This is an impossible result since it is more than the acid put into the solution, 0.100 M. Equation (3-2) (with **OH** negligible) gives

$$\mathbf{H} = 0.2(0.100 - \mathbf{H})/\mathbf{H}, \qquad \mathbf{H}^2 + 0.2\mathbf{H} - 0.02 = 0$$

$$\mathbf{H} = 0.075 \qquad \text{and} \qquad \mathrm{p}\mathbf{H} = 1.12$$

The calculations for the conjugate base and buffer are left for an exercise. One must be alert to results which make no sense and/or do not check in the full equation (3-2).

Figure 3-2 shows the effects of dilution upon pure solutions of weak acids of pK_a from 0 to 12, and the strong acid limiting line. A mirror image set above pH 7 would describe bases. For the strong acid line, we use the relation obtained from equation (3-2) by letting $K_a \rightarrow \infty$:

$$\mathbf{H} = C_a + \mathbf{OH} \qquad \text{(strong acids)}$$

Only near p**H** 7 does the **OH** term become important and produce a curve.

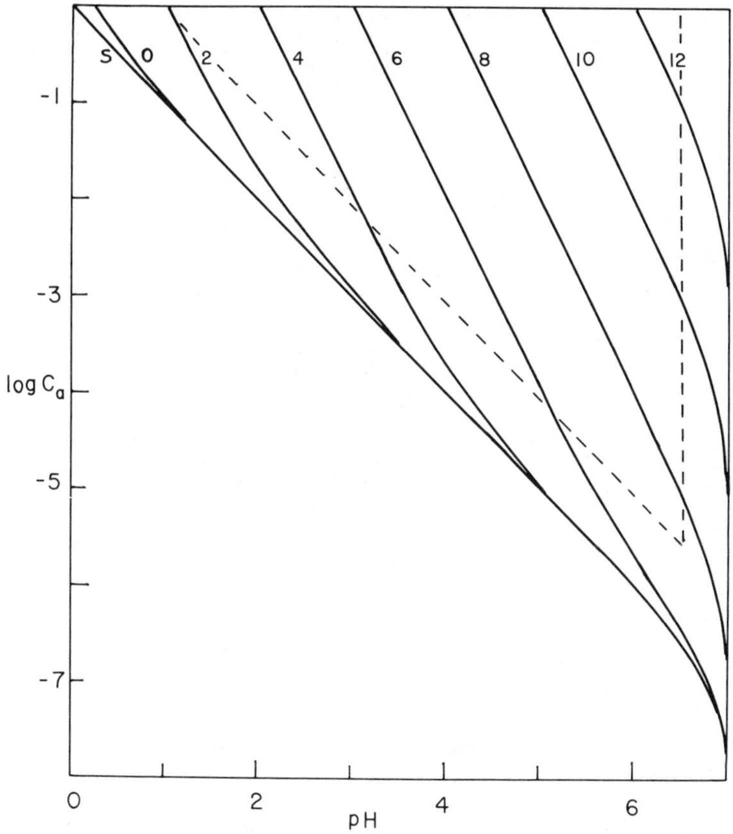

Figure 3-2. the **pH** of acids of various strengths as a function of concentration C_a. The pK_a of acids is indicated. The left limiting line is for the strong acid case. The dashed lines enclose area in which the approximate equation (3-3) is valid to 10% or better.

The limiting, left, line for a strong acid is joined by the weaker acids as they are diluted. The weak acid lines become straight at higher concentrations, where they follow equations (3-3)–(3-6). Note that acids having K_a at 1 and 10^{-2} (pK_a 0 and 2) do not obey the approximate equation at any concentration on this graph.

Study of these graphs should clarify the statement, "the approximate equations (3-3)–(3-6) apply if the acid is neither too dilute nor too strong." Students have usually been left without clear means to tell what *is* too dilute or too strong. We give two ways to know for sure: First, check the result in equation (3-2), and second, look at Figures 3-1–3-3.

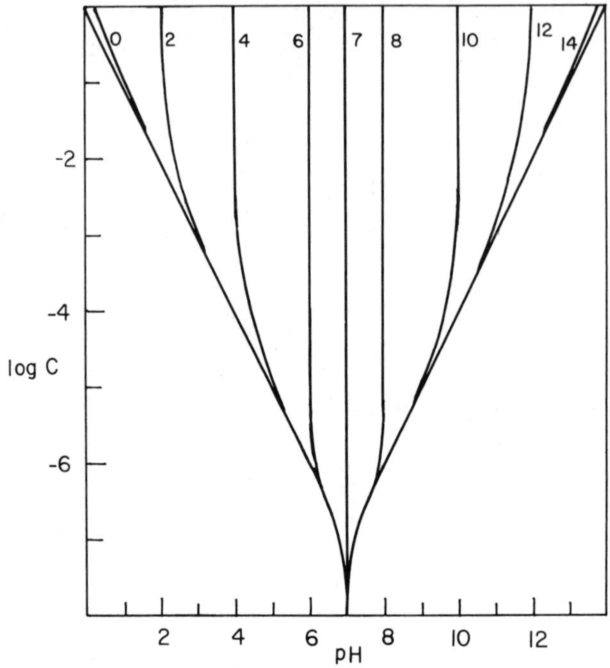

Figure 3-3. The variation of **pH** of buffers upon dilution. Each has equal analytical concentration of acid and conjugate base, $C_a = C_b$. The pK_a values are indicated. The strong acid and base cases are shown as limiting lines at each side.

If one wanted to enclose an area in Figure 3-2 in which the approximate equation (3-3) is valid to 10% for **H**, one might put in the line at which $C_a = 90K_a$, since at 10% ionic form we have

$$\mathbf{HX} \longrightarrow \mathbf{H} + \mathbf{X}$$

at equilibrium: $0.9C_a$ $\quad 0.1C_a$ $\quad 0.1C_a$

$$K_a = \frac{(0.1C_a)(0.1C_a)}{0.9C_a} \qquad \text{or} \qquad C_a = 90K_a$$

This is valid unless the **OH** term becomes important, which happens at about **pH** 6.5. So we also add that line. Any other level, say 1%, might be chosen as well. The idea is to show graphically the relation of K_a and dilution to equation (3-3). The identical situation applies with respect to **OH** and bases. Equation (3-4) can be rearranged by using $K_w = \mathbf{H}\cdot\mathbf{OH} = K_aK_b$, for any conjugate pair:

$$\mathbf{OH} = (K_bC_b)^{1/2}$$

This has the same form as equation (3-3). We could simply change the coordinates of Figure 3-2 to **pOH** and log C_b to make the graph for base solutions.

When known concentrations of conjugate acid and base are mixed, we have buffer solutions. Figure 3-3 shows the vertical straight lines where equation (3-5) applies. At the curved portions, equation (3-2) is required. The validity of equation (3-5) can also be seen in Figure 3-1, curve *b*.

While Figure 3-2 does fairly depict the full range of possibilities with acid (or base) solutions alone, Figure 3-3 shows only the 1:1 buffers. A full series of buffers ranging from pure acid to pure base can be shown by the plot in Figure 3-4, in which **pH** has been calculated from equation (3-2) for various values of C_a and C_b, holding their total constant at 0.100 *M*, except for one case at 0.001 *M* to show the dilution effect. Thus, we go from pure 0.1 *M* **HX** at the left to pure 0.1 *M* **X** at the right, with 0.05 *M* **HX**–0.05 *M* **X** at the midpoint fraction, 0.5.

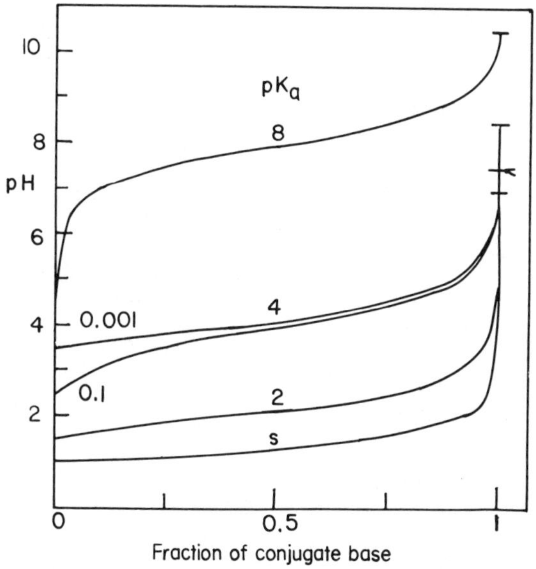

Figure 3-4. The variation of **pH** with change in the relative fractions of conjugate acid and base for total $C_a + C_b = 0.100$ *M*. $C_a + C_b = 0.001$ *M* is also shown for pK_a 4. The **pH** of the pure solutions of conjugate base are indicated by the horizontal lines at the right. Curves 2 and 4 both end at **pH** 7.5.

This also represents a constant-volume titration curve, showing **pH** at various degrees of neutralization of the acid by a strong base. Its form points up the meaning of buffering action. Good buffering occurs in the flatter portions where the buffer ratio changes only slowly with added acid or base. To clarify this, note the effect of a 10 % shift in C_b in two regions:

(a) From 1 % to 11 %: C_a/C_b goes from 99/1 to 89/11, or 99 to 8.1.
(b) From 40 % to 50 %: C_a/C_b goes from 60/40 to 50/50, or 1.5 to 1.0.

Remember that it is this buffer ratio which controls the **pH** when HX is weak [equations (3-2) and (3-5)]

$$\mathbf{H} \cong K_a C_a/C_b = K_a R, \qquad \mathbf{pH} = \mathrm{p}K_a + \mathrm{p}R \tag{3-7}$$

Examination of the methods of deriving the points for Figure 3-4 can exemplify all the common monoprotic acid cases. Let us look at them.

At the start: For pure HX, use equation (3-2) or (3-3) as required.

$$C_a = 0.10\,M \text{ [(3-3)]} \qquad \mathbf{H} = (K_a C_a)^{1/2}$$

$$\text{For } \mathrm{p}K_a = 8, \qquad \mathbf{H} = 10^{-4.5}, \quad \mathbf{pH} = 4.5$$

$$\mathrm{p}K_a = 4, \qquad \mathbf{H} = 10^{-2.5}, \quad \mathbf{pH} = 2.5$$

$$\mathrm{p}K_a = 2, \qquad \mathbf{H} = 10^{-1.5}, \quad \mathbf{pH} = 1.5$$

$$(\text{strong}, \qquad \mathbf{H} = 0.10\,M, \quad \mathbf{pH} = 1.0)$$

When we check the weak acids in equation (3-2) and look at Figure 3-2, we see that only for the strongest, $\mathrm{p}K_a = 2$, do we need a better answer. The full equation (3-2) gives **pH** = 1.55 for this case. This occurs again for the more diluted case, 0.0010 M, at $\mathrm{p}K_a = 4$. Here the approximate equation (3-6) gives us $\mathbf{pH} = \frac{1}{2}(4 + 3) = 3.5$. The full equation (3-2) gives 3.57. Note well that these two cases fall on the curved portion of Figure 3-2.

Along the curve: We have values for C_a *and* C_b, buffer solutions. The approximate equation (3-7) gives (for the $\mathrm{p}K_a = 4$ example, 0.1 M), at C_b 0.04, C_a 0.06,

$$\mathbf{pH} = 4.00 + \left(-\log\frac{0.06}{0.04}\right) = 4.00 - 0.18$$

$$\mathbf{pH} = 3.82$$

This result for **H** is under 1% of the 0.06 and 0.04, to which it is added and subtracted in the full equation (3-2). Thus, this answer will be found satisfactory.

At $C_b = C_a = 0.05$, we can see on Figure 3-3 that the 0.1 M cases for pK_a 4 and 8 will be correct with the approximate equation (3-5), pK_a = **pH**. For the other two cases again equation (3-2) will be required. Several other points in the first and last 10% of the curves should be obtained to establish the line.

At the end: We have pure 0.100 M conjugate base solution and can try equation (3-4),

$$\mathbf{pH} = \tfrac{1}{2}(\mathrm{p}K_a + \mathrm{p}K_w - \mathrm{p}C_b) = 8.5$$

for the pK_a 4 case.

Note that in every case where the full equation (3-2) has been needed, this can be predicted from examination of Figures 3-2 and 3-3. Furthermore, only when the **pH** is within about 0.5 unit of 7.0 is the full cubic form required. For most cases, either **H** or **OH** is negligible for the addition or subtraction, and a quadratic equation is quite good enough. Experience will be gained by practice.

3. Procedure for Solving pH Problems

Now that we have examined the consequences of the mathematical relations required by equilibrium, let us look at some applications to the kinds of solutions commonly used.

A prime requirement for all our calculations must be stressed at the start: Solutions must be at equilibrium in order for calculations with equilibrium constants to be valid. This is obvious, but can cause great difficulty because chemical knowledge must be brought into play to decide whether a chemically stable solution is present. For example, how does one treat the acidity of a mixture which is made so that it will be 0.1 M in hydrochloric acid and 0.2 M in sodium hydroxide? No such solution exists. One must know that these are strong electrolytes and that such high concentrations of **H** and **OH** react to form water. Thus, the mixture given turns into 0.1 M Na^+, Cl^- and 0.1 M Na^+, OH^-, that is, half the base is neutralized by the acid present. The approximate **pH** is 13. We do not follow the procedure of one canny student who replied that the **pH** of the 0.1 M acid is 1.0 and that of the 0.2 M base is 13.3, and thus the average is 7.15.

Example 3. Estimate the **pH** and the pa_H [pa_H = p(H^+_{aq})] of a mixture made by placing 50 ml of 6.0 M acetic acid and 50 ml of 1.00 M NaOH in a liter volumetric flask and diluting it to 1000 ml.

Step 1. The weak acid and strong base must react before a stable solution can be considered. We are given (50 ml)(6 mmol/ml) = 300 mmol of acetic acid, and (50 ml)(1 mmol/ml) = 50 mmol of NaOH. Fifty millimoles of each react and 250 of the acid is left in excess

$$\begin{array}{lccccc} \text{at start:} & 300 & & 50 & & \text{(in millimoles)} \\ & HX & + & OH^- & \rightarrow & HOH + X^- \\ \text{after equilibrium:} & 250+x & & x & & 50-x \end{array}$$

After hypothetical complete neutralization, some, x, must return to establish equilibrium since **OH** cannot be zero. (H_2O remains near 55 M.)

Step 2. Now that we have a chemically stable solution, we decide what type of situation we have and what mathematical approach is suitable. The analytical concentrations for the resulting solution are

$$C_a = 250 \text{ mmol}/1000 \text{ ml} = 0.250\ M$$

$$C_b = 50 \text{ mmol}/1000 \text{ ml} = 0.050\ M$$

Since we have known values for C_a and C_b, this is an acetate buffer. For an approximate solution, try equation (3-7) with pK_a° = 4.76 (zero ionic strength):

$$\mathbf{pH} \cong pK_a + pR = 4.76 + \left(-\log\frac{0.250}{0.050}\right) = 4.06$$

Step 3. Now, check this result in equation (3-2), using the right side:

$$\mathbf{H} = 10^{-4.76}\frac{(0.250 - 10^{-4.06} + 10^{-9.94})}{(0.050 + 10^{-4.06} - 10^{-9.94})} = 10^{-4.06}$$

Thus, **H** and **OH** are truly negligible on the right, and it becomes equations (3-5) and (3-7).

Step 4. For a more correct result, we look up, in the tables of Appendix A-1, or calculate [Chapter 2, equation (3)] activity coefficients for the conditional K_a:

$$I = \tfrac{1}{2}cz^2 = \tfrac{1}{2}([Na^+] + [C_2H_3O_2^-]) = 0.05\ M$$

We omit the small **H** from the ion sum as negligible here. This gives us

K_a for these conditions as 2.6×10^{-5}. Thus,

$$\mathbf{H} = 2.6 \times 10^{-5}\left(\frac{0.250}{0.050}\right) = 1.3 \times 10^{-4}$$

The pH meter reading of $-\log(H_3O^+)$, using the Kielland f_+ of 0.85, yields

$$pa_H = -\log(1.3 \times 10^{-4} \times 0.85) = 3.96$$

Example 4. How many milliliters of 1.00 *M* HCl solution should be added to 100.0 ml of 0.100 *M* NH_3 to produce a buffer of pa_H 9.00? A chemical reaction is implied. We follow the same steps as in Example 3.

Step 1. We have 10.0 mmol of ammonia and *x* mmol of HCl, since *x* ml of 1 *M* is *x* mmol:

$$\begin{array}{lccccc} \text{at start:} & x & x & 10 & & \\ & H_3O^+ + & Cl^- + & NH_3 & \longrightarrow NH_4^+ + & Cl^- \\ \text{at equilibrium:} & 10^{-9}\,M & x & 10-x & x & x \end{array}$$

Step 2. For an approximate answer, take the hydronium activity to be the molarity and use the K_a° for the ammonium system, using equation (3-5):

$$K_a \cong 10^{-9.24} = \frac{\mathbf{H}[NH_3]}{[NH_4^+]} = \frac{10^{-9.00}(10 - x)/v}{x/v}$$

(We divide mmol by the final volume v to get molarities. Here, as often happens, we are lucky and v cancels). Solve for x:

$$x = 10^{0.24}(10 - x), \qquad x = 6.35 \text{ ml of } 1\,M \text{ HCl}$$

Step 3. Final volume v is 100 + 6.35 ml, so we can find the molarities:

$$C_a = 6.35 \text{ mmol}/106.35 \text{ ml} = 0.0596\,M\ NH_4^+$$

$$C_b = 3.65 \text{ mmol}/106.35 \text{ ml} = 0.0343\,M\ NH_3$$

Figure 3-3 shows us that this buffer (of *R* not too far from 1) at pK_a 9.24 is likely well within the region covered by the approximate equation (3-5) or (3-7). In addition, we can see in equation (3-2) that our **H**, 10^{-9}, and our **OH**, 10^{-5}, are negligible with respect to C_a and C_b. So we need only check the ionic strength and make proper activity corrections to obtain valid results.

Step 4. $I = \frac{1}{2}([NH_4^+] + [Cl^-]) = 0.06\ M$, other ions being negligible. Since the hydronium required is stated as an activity, we need only f_+ for the NH_4^+ ion. Taking the activity coefficient from the Kielland Table A-1 of Appendix A-1, we obtain

$$K_a^\circ = 10^{-9.24} = \frac{(H_3O^+)(NH_3)}{(NH_4^+)} = \frac{10^{-9.00}[NH_3]}{[NH_4^+]0.79}$$

Returning to step 2, we get

$$x = 2.20(10 - x), \qquad x = 6.88 \text{ ml HCl}$$

The revised concentrations are

$$C_a = 6.88/106.88 = 0.0645\ M\ NH_4^+$$

$$C_b = 3.12/106.88 = 0.0292\ M\ NH_3$$

The slight change in ionic strength does not warrant recalculation, because of the precision limit in the activity coefficients available.

Example 5. *A different charge type.* Find the approximate **H** and pa_H of a buffer having 0.020 M $NaHSO_4$ and 0.020 M Na_2SO_4. The HSO_4^- is a monoprotic acid of $pK_a^\circ = 1.99$.

Steps 1 and 2. First from equation (3-5), using $[HSO_4^-] = [SO_4^{2-}] = 0.020\ M$,

$$K_a = 10^{-1.99} = \frac{\mathbf{H}[SO_4^{2-}]}{[HSO_4^-]}$$

$$\mathbf{H} = 10^{-1.99} = 1.02 \times 10^{-2}$$

Step 3. We are on a curve in Figure 3-3 and the **H** term is large in equation (3-2),

$$\mathbf{H} = \frac{10^{-1.99}(0.020 - \mathbf{H})}{(0.020 + \mathbf{H})}$$

The **OH** is 10^{-12} (negligible),

$$\mathbf{H}^2 + 0.0302\mathbf{H} - 2.04 \times 10^{-4} = 0$$

and

$$\mathbf{H} = 0.0057$$

which is the approximate acidity of the solution.

Step 4. Activity correction: The ions in this solution are (using step 3)

$[Na^+]$	$0.060\ M$	$[HSO_4^-]$	$0.0143\ M$
H	$0.0057\ M$	$[SO_4^{2-}]$	$0.0257\ M$

and

$$I = \tfrac{1}{2}(cz^2) = \tfrac{1}{2}(0.060 + 0.0143 + 0.0257 \times 4 + 0.006) = 0.0915\ M$$

Notice the z^2 term is 4 for SO_4^{2-} ion and 1 for the rest. We now look up the f values in the Kielland table in Appendix A-1 and get K_a,

$$K_a = \frac{K_a^\circ f_-}{f_+ f_{2-}} = \frac{\mathbf{H}[SO_4^{2-}]}{[HSO_4^-]} = \frac{1.02 \times 10^{-2}(0.78)}{(0.83)(0.38)} = 2.5_2 \times 10^{-2}$$

Repetition of step 3 with this corrected value gives $\mathbf{H} = 0.009_2$. This multiplied by 0.83 to get a_H yields $pa_H = 2.12$. The new I is $0.098\ M$. (An improved result can be obtained by recalculation with this value).

4. Summary

After these examples, it should be clear that a definite pattern can be seen in the steps used to solve these problems. Let us summarize them and emphasize certain troublesome points.

1. Find effective analytical concentrations to use, C_a and/or C_b: To do this, we must decide whether the material given should undergo extensive neutralizations. These must be allowed to go to completion to find the effective C_a and C_b for the calculation. To make this decision, one must know which ions associate into weak electrolytes. Tables of equilibrium constants may need to be consulted. Experience and a few rules can help.

(a) Most acids are weak. HX is strong only for $X = Cl^-$, Br^-, I^-, SCN^-, NO_3^-, ClO_4^-, HSO_4^-, and a few others. That means we do not write these ions in protonic equilibria in water solutions. Other anions do react,

$$X^- + H_2O \rightleftharpoons HX + OH^-$$

for the overwhelming majority of anions, acetate, fluoride, carbonate, etc. See Figure 1-4.

(b) Soluble metal hydroxides are strong: NaOH, LiOH, KOH, $Ba(OH)_2$. However, some weak ion association occurs with the multiply charged ions. Tables must be consulted for accurate work. Most other bases are weak: ammonia, amines, and the anions not listed in (a).

(c) Other soluble ionic compounds (old term "salts") are assumed to be strong, but there are many exceptions at the level of weak ion association or complexing. Tables must be consulted.

2. Write the chemical equation for the equilibrium involving the conjugate acid and base. Write the K_a° expression. Put into this the values known, and x for the unknown quantity desired. Solve roughly for x using all f values equal to 1. Use approximate equations (3-3), (3-4), or (3-5) for acid, base, or buffer mixtures.

3. Look at Figure 3-2 or 3-3 and equation (3-2) to see if the result from step 2 is valid. Solve anew with the quadratic or cubic equation when needed.

4. Add the ions by equation (2-4) to obtain the ionic strength. If this is above 0.001 for singly charged cases, or above 10^{-5} for multiply charged ions, make activity coefficient corrections and obtain a refined result.

Problems

1. Consider mixing 10.0 ml of each solution in the following groups. Decide what, if any, neutralizations should occur. Calculate the effective C_a and/or C_b of the solution, remembering to take into account the dilution upon mixing, which doubles or triples the volume. State whether each case is that of a weak acid, buffer, or a weak base, or whether an excess of strong acid or base is left to predominate in setting the **pH**.
   a. 1.0 *M* NaOH and 2.0 *M* acetic acid.
   b. 1.0 *M* NH_3, 0.20 *M* NH_4Cl, and 0.10 *M* HNO_3.
   c. 1.00 *M* H_2SO_4 and 1.40 *M* NaOH.
   d. 2.0 *M* NaOH and 1.0 *M* acetic acid.

2. Calculate rough **pH** values for each solution mixed in Problem 1.

3. Prove the relation $K_aK_b = K_w$ by multiplying the equilibrium constant expressions for the conjugates in each of the following cases.
   a. Acetic acid–acetate. b. Ammonium ion–ammonia.

4. Demonstrate that equation (3-2) is valid for the different charge types by deriving it from the charge and material balances for the following cases.
   a. NH_4^+–NH_3 systems. b. HSO_4^-–SO_4^{2-} systems.

5. Calculate **H** for the 1:1 buffer and the pure conjugate base, all at analytical concentrations of 0.100 *M*, for the trichloroacetate system begun in this chapter, Example 2. Check the result on Figure 3-1, *b* and *c*. Take the conditional K_a as 0.20, although it is not well established.

6. Calculate and compare the **pH** of 10^{-7} *M* NaOH and NH_3 solutions. You can check the results on Figure 3-2 by noting that it is valid if you change pK_a to pK_b and **pH** to **pOH**. The pK_b for NH_3 is 4.7.

7. Calculate the **H** of a buffer made with analytical concentrations 0.200 *M* lactic acid and 0.100 *M* sodium lactate. Repeat after a 100-fold dilution of the solution and compare the need for the full equation (3-2) in each case. Take a constant ionic strength of 0.10 *M* in both cases and use *f* values from the Kielland table in Appendix A-1, assuming lactate ion is about the same size as acetate. The reported pK_a° for lactic acid is 3.858. It is $CH_3CH(OH)COOH$.

---

8. A solution of pure pyruvic acid [$CH_2{=}C(OH)COOH$] was titrated with pure NaOH solution at the pH meter. Half-way to the equivalence point the meter reading was 2.555. The analytical concentrations calculated from the starting molarities were each 0.0600 *M*, which is also the ionic strength. Calculate K_a°.

9. What are the changes in log units in going from pK_a° to the pK_a at 0.10 *M* ionic strength for acetic acid, NH_4^+, and for HSO_4^-? What are the changes for the corresponding pK_b values?

10. The pK_a° reported for glycollic acid is 3.882 at 25°. What is the **pH** at the start, midpoint, and end when 25.00 ml of 0.0500 *M* glycollic acid is titrated with 0.2000 *M* NaOH solution? First calculate the effective C_a and/or C_b using the new volume at each point. Then get an approximate **H**. How large will the activity effect be on the K_a at the conditions, and will equation (3-2) be required to get a result to ± 0.01 pH unit?

4 Monoprotic Acid–Base Diagrams

In this chapter we introduce a variety of graphical displays of monoprotic acid–base equilibria. Although a number of diagrammatic methods for solving pH problems has been proposed, none is as fast as numerical approximation using equations (3-2)–(3-5), for those who are experienced in chemistry. For the student, however, these diagrams can help develop the experience needed to make correct approximations suitable under various conditions. First, we present a complete ratio diagram since it is the simplest to use and the most foolproof to interpret. The log α and log C diagrams shown next are either more complicated (log α) or less foolproof, requiring estimations as to the approximate equalities or intersections to look for (log C).

1. Equilibrium Ratio Plots

Here we shift our attention from the concentration of each species in equilibrium to the ratio of acid to conjugate base at equilibrium, since this ratio fixes the **pH**. From the K_a expression,

$$\mathbf{H}/K_a = \mathbf{HX}/\mathbf{X} = R_e, \qquad \log R_e = \mathrm{p}K_a - \mathbf{pH}$$

This equilibrium ratio must not be confused with the buffer ratio of equation (3-7), which is C_a/C_b, the ratio of the analytical concentrations, which may or may not be close to the ratio of the species at equilibrium. The equilibrium ratio is given by $\mathbf{H}/K_a$. A material balance expression for this ratio can be found from the complete

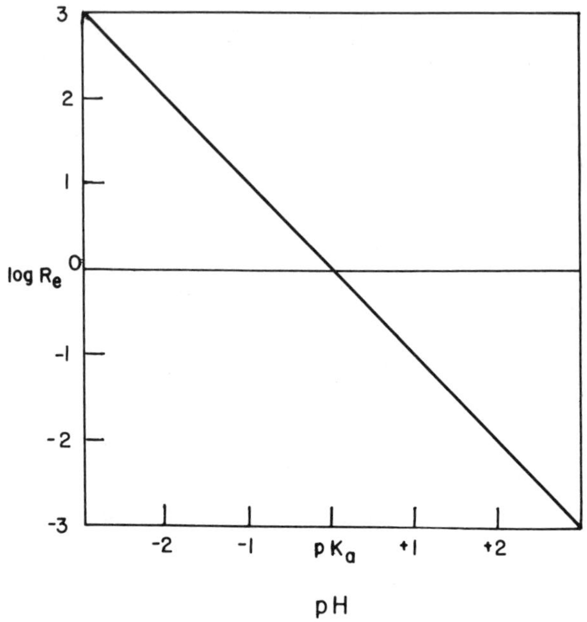

Figure 4-1. The logarithm of the equilibrium ratio of conjugates for any monoprotic system. **pH** measured from pK_a value at center.

equation (3-2),

$$\mathbf{H}/K_a = (C_a - \mathbf{H} + \mathbf{OH})/(C_b + \mathbf{H} - \mathbf{OH}) = R_e \qquad (4\text{-}1)$$

Note that this becomes the buffer ratio C_a/C_b only when both **H** and **OH** are negligible. The plot of the log of $\mathbf{H}/K_a$ (log R_e) is a straight line as shown in Figure 4-1, crossing the log $R_e = 0$ line at $\mathbf{pH} = \mathrm{p}K_a$. Thus, it can be quickly constructed for any chosen acid–base system. The material balance (right) side of equation (4-1) depends upon the concentrations chosen for C_a and C_b. Either may be zero. This method is not limited to buffers. The intersection of these two plots gives the **pH** of the solution. This also suggests a simple method of programming a computer to solve **pH** problems for these systems.[1]

Figure 4-2 shows the material balance R_e plots for 0.100 M solution of three cases: (a) pure acid HX; (b) pure base, $C_b = 0.100\ M$, $C_a = 0$; and (c) buffer, $C_a = C_b = 0.100\ M$. Note that this function is independent of any particular acid–base chosen. It only tells the

[1]See C. H. LANGFORD, *BASIC Equilibrium Calculations*, Addison-Wesley, Reading, Massachusetts, 1973.

proportions of proton binding or donating that must occur if a certain **pH** is to exist with a chosen set of values for C_a and C_b. For example, to obtain **pH** 4 with 0.1 M HX ($C_b = 0$), the R_e on the material balance side of equation (4-1) is

$$R_e = (0.1 - 0.0001)/0.0001 = 999$$

The **OH** is negligible here, and the log R_e is thus near 3. To have **pH** 1, all the HX must form ions and R_e is zero, log $R_e \to -\infty$. We have taken ionic strength 0.1 M throughout this section, so that pK_w is 13.80. Thus, a neutral solution occurs at **pH** 6.9 where **H** = **OH** = $10^{-6.90}$. This is where the denominator of our R_e fraction goes to zero for the pure acid case and log $R_e \to \infty$.

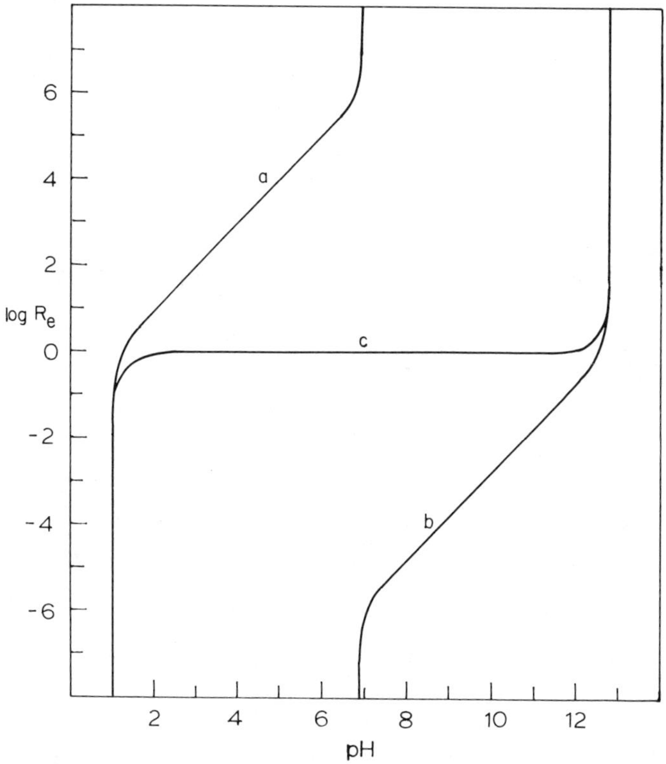

Figure 4-2. The logarithm of the ratio of conjugates required by the material balance condition for 0.100 M solutions of (a) pure acid, (b) pure base, (c) 1:1 buffer, $C_a = C_b = 0.100$ M. Assume 0.1 M ionic strength, pK_w = 13.80, neutral **pH**, 6.90.

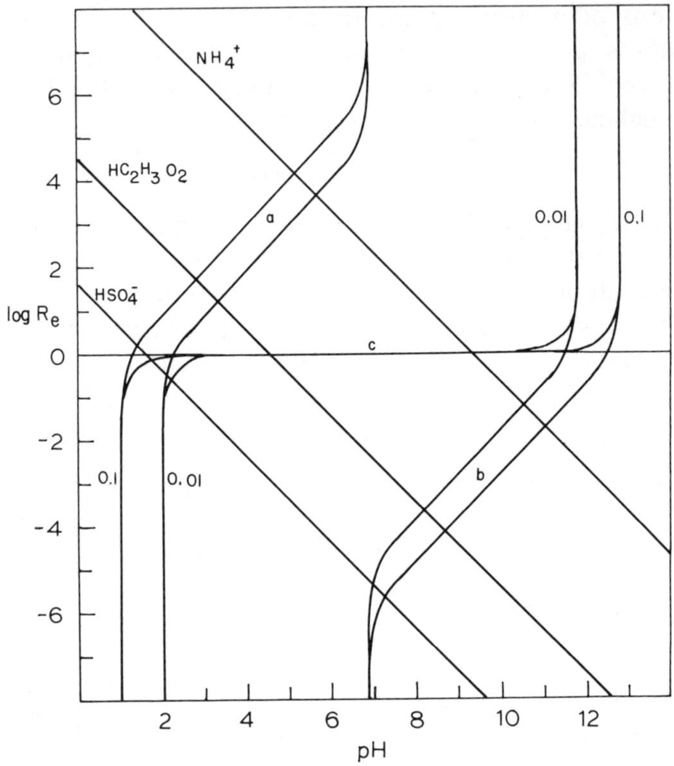

Figure 4-3. **pH** calculation diagram combining Figures 4-1 and 4-2. The 0.0100 *M* material balance cases added. Three equilibrium K_a cases shown for the acids bisulfate, acetic, and ammonium at 0.1 *M* ionic strength values at 25°C.

For the base, a mirror image of the acid plot is found. For the buffers, **H** and **OH** are negligible in the central region, where these are both small, and R_e is effectively the same as the buffer ratio C_a/C_b. The curved portions show the behavior of rather strong acid or base, where equation (3-7) is not valid.

To find the **pH** of any 0.100 *M* solution of any acid, base, or 1:1 buffer, we simply put a straightedge at 45° on this plot passing through **pH** = pK_a of the acid–base as in Figure 4-1. For other concentrations and buffer ratios we shall have to plot the required material balance R_e curves. Six cases are shown in Figure 4-3: the three from Figure 4-2 and the three for 0.0100 *M* acid, base, and buffer. Three equilibrium condition ratio (**H**/K_a) lines are shown passing through their $-\log K_a$ values for 25° and 0.100 M ionic strength: HSO_4^-, 1.60; acetic acid, 4.56; and NH_4^+, 9.29. These are representative of a wide variation in acid–base strength. The correct values of **pH** are read off at the intersections. Only for the hydrogen sulfate cases are these different from the approximate equation results. Note that these are molarity **pH** values. The effects of concentration and buffer ratio are further illustrated in Figure 4-4. The same three acid–base $\mathbf{H}/K_a$ lines are included.

The pure acid and base material balance lines in Figure 4-2 resemble the corresponding lines of Figure 3-1. Examination of equations (3-2)–(3-6) will reveal why this is the case.

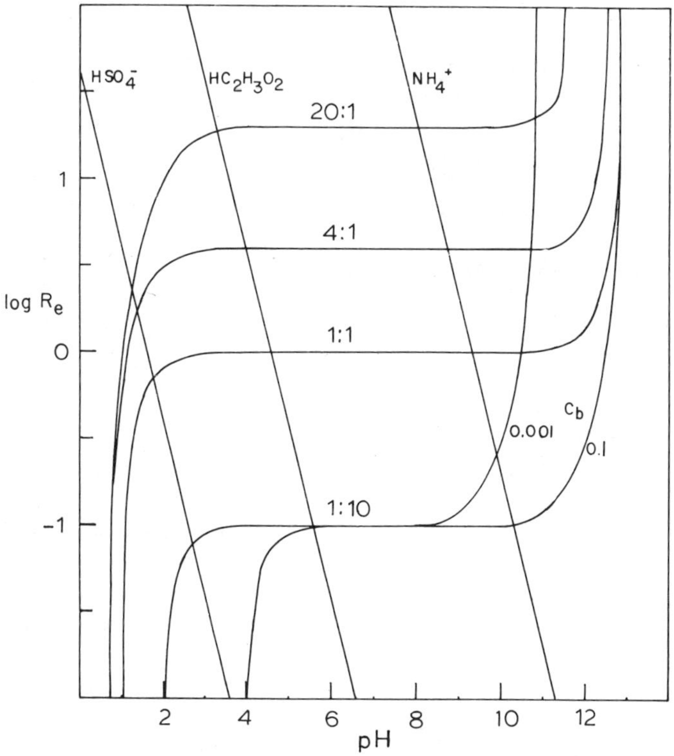

Figure 4-4. Ratio method **pH** calculator for buffers of various concentrations: 20:1, $C_a = 0.2\ M$, $C_b = 0.01\ M$. 4:1, $C_a = 0.2\ M$, $C_b = 0.05\ M$. 1:1, $C_a = C_b = 0.1\ M$, 1:10, $C_a = 0.01\ M$, $C_b = 0.1\ M$, and also $C_a = 10^{-4}\ M$, $C_b = 10^{-3}\ M$.

This ratio method is useful if one is to determine the **H** at a specific concentration, 0.1 or 0.01 *M*, of acid, base, or buffer. One needs only a straight edge to read off **pH** values. To obtain valid results, as with any method, one must use activity coefficients to convert K_a° and K_w° to their effective molarity values at the ionic strength given. To find pa_H, we simply add the $\log f_+$ to the **pH**. For example, for the acetate buffers in Figure 4-3, we read off **pH** 4.56 and add 0.08 to get 4.64, the correct pa_H read at the pH meter at 25°. (Our 0.01 *M* buffers were taken to be in the presence of salt to give ionic strength 0.1 *M*. For example, 0.09 *M* NaCl added to the 0.01 *M* $NaC_2H_3O_2$ gives us ionic strength 0.1 *M*.)

2. Logarithmic Fraction and Concentration Diagrams

The equilibrium fractions of species are closely related to their ratio, so that it is not surprising that the fraction function has been used in the way we have used R_e above. The fractions are related to the equilibrium constant and **H**:

$$\begin{aligned} \alpha_0 &= \mathbf{X}/(\mathbf{X} + \mathbf{HX}) = 1/(1 + \mathbf{HX}/\mathbf{X}) = 1/(1 + R_e) \\ \alpha_1 &= \mathbf{HX}/(\mathbf{HX} + \mathbf{X}) = 1/(1 + \mathbf{X}/\mathbf{HX}) = 1/(1 + R_e^{-1}) \end{aligned} \tag{4-2}$$

Substituting from the K_a expression, $\mathbf{H}/K_a$ for R_e, we get the relation of the fractions α_0 and α_1 to **H**:

$$\alpha_0 = (1 + \mathbf{H}/K_a)^{-1}, \qquad \alpha_1 = (1 + K_a/\mathbf{H})^{-1} \tag{4-3}$$

Let us plot a general graph of these fractions, Figure 4-5. Equation (4-3)

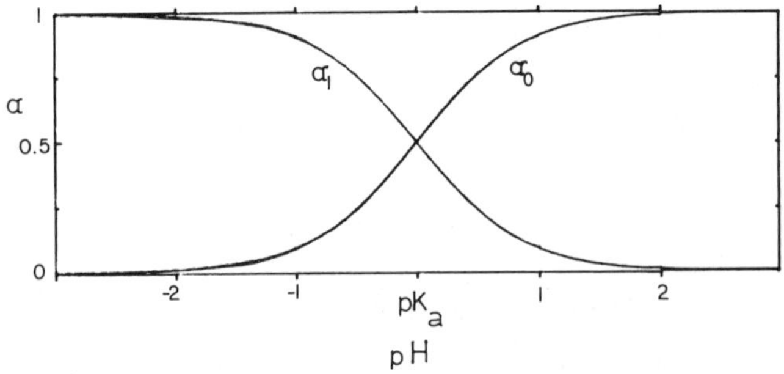

Figure 4-5. General monoprotic equilibrium α fractions vs. **pH**.

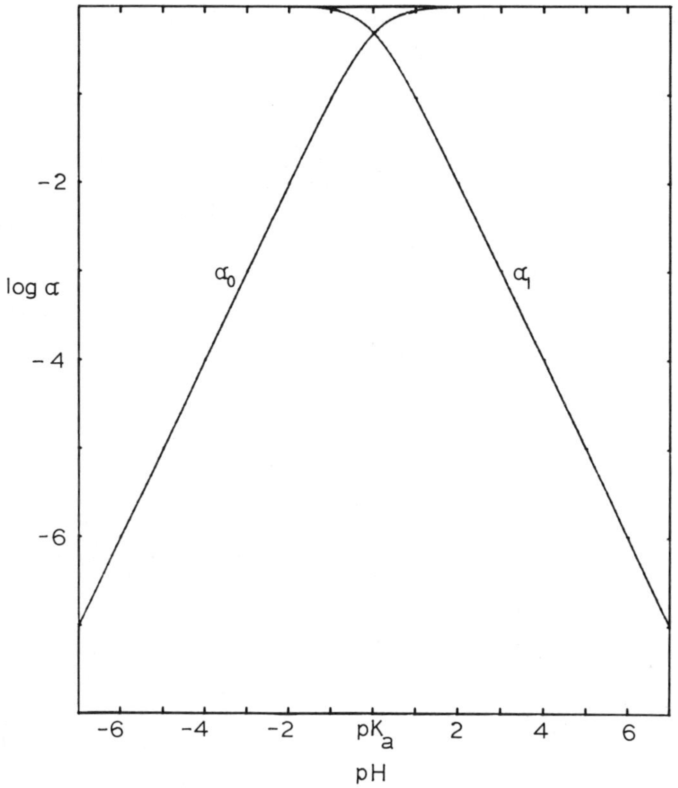

Figure 4-6. Logarithmic α fraction curves. 14 **pH** unit spread.

yields the following results:

			pH
$\mathbf{H} = K_a$,	$\alpha_0 = \alpha_1 = 0.5$		$\mathrm{p}K_a$
$\mathbf{H} = 10K_a$,	$\alpha_0 = 0.09$,	$\alpha_1 = 0.91$	$\mathrm{p}K_a - 1$
$\mathbf{H} = 100K_a$,	$\alpha_0 = 0.01$,	$\alpha_1 = 0.99$	$\mathrm{p}K_a - 2$
$\mathbf{H} = 0.1K_a$,	$\alpha_0 = 0.91$,	$\alpha_1 = 0.09$	$\mathrm{p}K_a + 1$

etc. To obtain the curves for acetic acid at 0.1 *M* ionic strength, for example, use 4.56 for pK_a in this process. Figure 4-5 shows how the fraction α_0 rises with **pH** as α_1 falls, the fraction of HX decreases and that of X^- rises as strong base is added to convert HX to X^-. At all points, $\alpha_0 + \alpha_1 = 1$. More generally useful will be the plots of log α shown in Figure 4-6. Here one can read the smaller values which look like zero in Figure 4-5.

To construct a pH problem-solving method we proceed just as with the ratios and find the material balance expression for the fractions just derived from the equilibrium condition equation. From the derivation of equation (3-2) we have (C is the total of **HX** and $\mathbf{X}^-$)

$$\alpha_0' = \mathbf{X}/C = (C_b + \mathbf{H} - \mathbf{OH})/(C_a + C_b)$$
$$\alpha_1' = \mathbf{HX}/C = (C_a - \mathbf{H} + \mathbf{OH})/(C_a + C_b) \quad (4\text{-}4)$$

We have called these material balance functions the α' fractions. They equal the equilibrium α fractions only at one value of **H** at which equilibrium and material balance conditions are simultaneously satisfied. This can be found graphically or programmed for computer solution. For pure acid and base, C_b or C_a zero, a simple, invariant shaped curve results in the logarithmic plots.

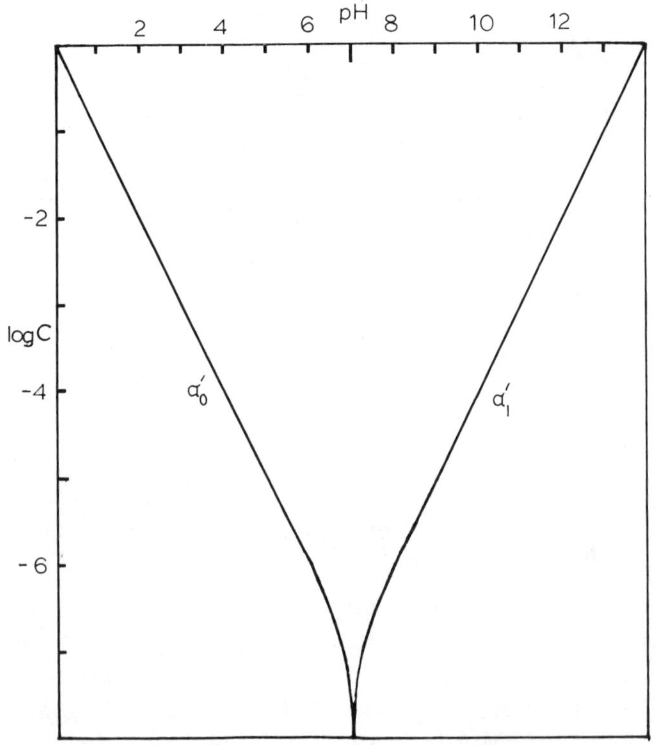

Figure 4-7. Material balance curves. For 1 M pure acids, $\log \alpha_0' = \log \mathbf{X}$. For 1 M pure bases, $\log \alpha_1' = \log \mathbf{HX}$. Log C refers to **X** on left and to **HX** on right.

For pure acid solutions, with $C_b = 0$, we get from equation (4-4)

$$\alpha'_0 C_a = \mathbf{H} - \mathbf{OH} = \mathbf{D}, \qquad \log \alpha'_0 = \log \mathbf{D} - \log C_a \qquad (4\text{-}5)$$

The plot of $\log \alpha'_0$ vs. $\mathbf{pH}$ is a straight line except near pH 7, where **OH** is not much smaller than **H**. This line is shown as the left half of Figure 4-7 up to $C_a = 1$. These are the α's for $C = 1$. For pure base solutions, with $C_a = 0$, we get from equation (4-4)

$$\alpha'_1 C_b = \mathbf{OH} - \mathbf{H} = -\mathbf{D}, \qquad \log \alpha'_1 = \log(-\mathbf{D}) - \log C_b \ (4\text{-}6)$$

This produces a mirror-image plot of the acid case above, and is shown as the right half of Figure 4-7. The log C shown as ordinate refers to $\log C_b\alpha'_1$ or $\log C_a\alpha'_0$, which are the logs of the concentrations of ions produced in the solution from either C_a or C_b. This method is used to help in the following problem-solving method. This log C is also log α for the cases C_a or C_b at 1 M. Thus, this same set of curves serves for all acid or base solutions, but not for buffers. We are now ready to find intersections of the equilibrium and material balance α curves to solve problems.

Example 1. Find the **pH** of 0.010 M acetic acid, and also of 0.010 M sodium acetate.

When we place the top lines of the equilibrium α curves (Figure 4-6) on the α' curves of Figure 4-7 at $\log C = -2$ and with their intersection at $\mathbf{pH} = \mathrm{p}K_a = 4.56$ for total ionic strength 0.1 M, we produce the log C curves for **HX** and **X**, acetic acid molecules and acetate ions for the conditions of this problem. This superposition is shown in Figure 4-8. The intersections give the simultaneous solutions for the two cases, s_a for the acid at **pH** 3.28, and s_b for the base at **pH** 8.28. From the α lines, which are now log C lines, we can read off $[C_2H_3O_2^-] = 10^{-3.28}$ in the acetic acid solution, and $[HC_2H_3O_2] = 10^{-5.72} = \mathbf{OH}$ in the acetate solution.

Thus the approximate equations (3-3) and (3-4) hold here. This is not the case in the following example.

Example 2. Find the **pH** of 0.020 M $NaHSO_4$ and of 0.020 M Na_2SO_4 under conditions giving the conditional pK_a as 1.74.

We have $\log C = \log 0.020 = -1.70$. We place the curves of Figure 4-6 upon those of Figure 4-7 at log C, -1.70, and pK_a intersection at **pH** 1.74. This produces Figure 4-9, from which we read the solution intersections s_a at **pH** 1.90 for the acid and s_b at **pH** 7.15 for the

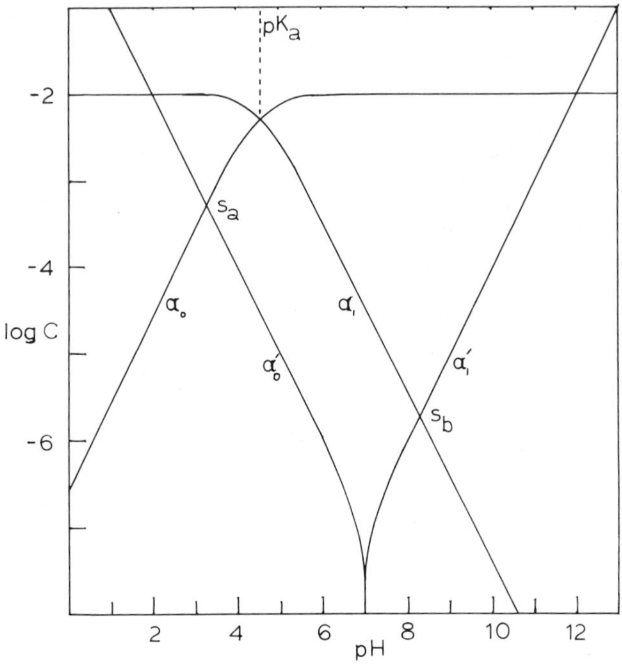

Figure 4-8. Combination of Figures 4-6 and 4-7 as a **pH** calculator for 0.01 *M* solutions of acetic acid (intersection s_a) and of sodium acetate (intersection s_b) at ionic strength 0.1 *M*.

base. This is a case in which the curvatures at the intersections show us that the approximate equations will not apply. The amount of HSO_4^- forming ions in the acid case is large. We read from the α lines that the $[HSO_4^-]$ at equilibrium is 0.008 and the $[SO_4^{2-}]$ is 0.012. In the base solution, however, α_1 shows us that only $10^{-7.2}$ *M* HSO_4^- forms in the 0.020 *M* SO_4^{2-} solution.

In this case, our sliding calculator saved solving quadratic equations. Such a sliding calculator can be constructed by putting the curves of Figures 4-6 and 4-7 on transparent plastic so that they can be superposed for any chosen pK_a and concentrations. The 14-unit spread of Figure 4-7 is correct only for pK_w 14.00. More generally, the V-curve of Figure 4-7 can be centered on $pK_w/2$ for any conditions and moved up so that $\log \alpha'_0$ and $\log \alpha'_1$ hit $\log C$ at zero and **pH** 0 and pK_w. For an example, we have published values for 25° and ionic strength 0.5 *M*: pK_w 13.68 and pK_a 4.48 for acetic acid. Place the top of Figure 4-7 to span **pH** 0–13.68 on the $\log C = 0$ line of graph paper,

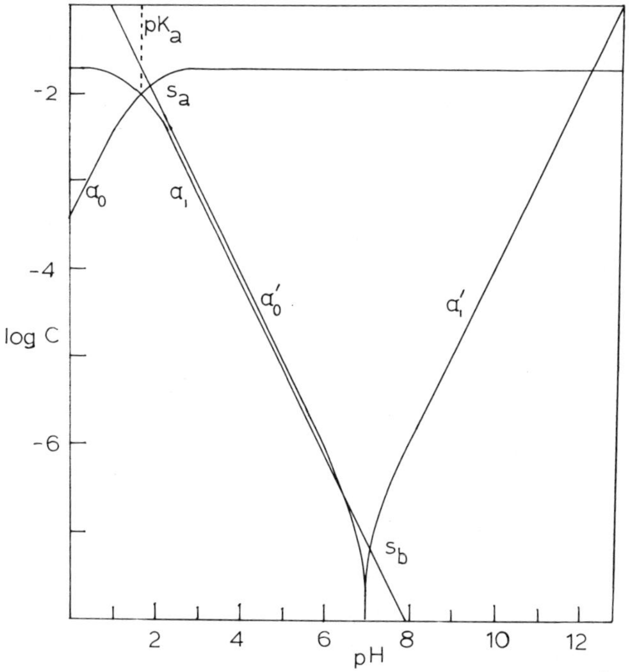

Figure 4-9. Log α intersections for 0.020 *M* $NaHSO_4$ (s_a) and for 0.020 *M* Na_2SO_4 (s_b).

and the Figure 4-6 lines centered on **pH** 4.48 at any chosen concentration. Note that only the pK_a change affects the **pH** of the pure acid solution, while the changes in both constants affect the **pH** of the conjugate base solutions. See equations (3-3) and (3-4).

Buffers require a different treatment since their α' curves do not have invariant shapes like those above. Using total $C = C_a + C_b$, we can write equations from (4-4), for buffers,

$$\alpha_0' = C_b/C + (\mathbf{H} - \mathbf{OH})/C, \qquad \alpha_1' = C_a/C - (\mathbf{H} - \mathbf{OH})/C \qquad (4\text{-}7)$$

Thus, to calculate α' values to plot, we must choose not only **H** but also C_a and C_b, and we get different curves for various concentrations. A set of α_1' curves is shown in Figure 4-10 for a variety of buffer ratios and total C constant at 0.100 *M*. When the buffer ratio R_b, which is C_a/C_b, is equal to infinity, we have $C_b = 0$, the pure 0.100 *M* HX solution. At the other extreme, $C_a = 0$, we have the pure conjugate base. These lines have been included for comparison. The log α_0' set

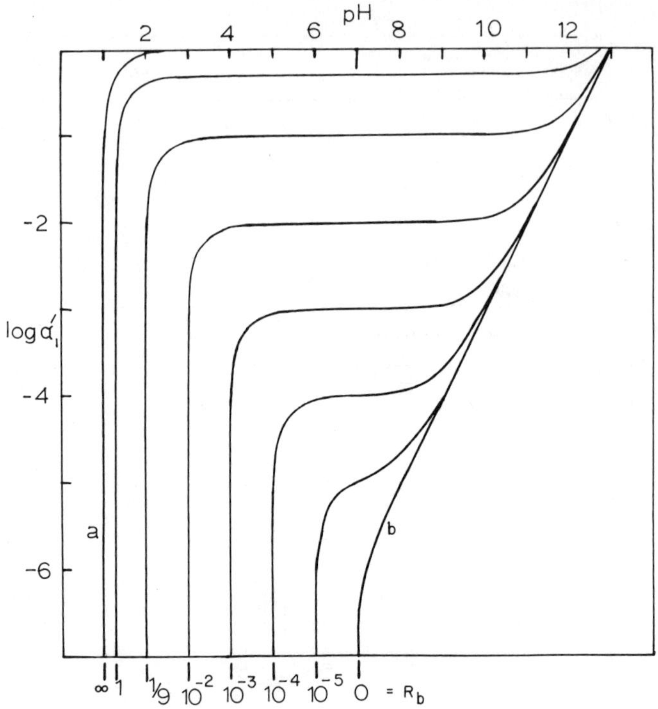

Figure 4-10. Log material balance fraction curves for various buffer ratios at constant total $C_a + C_b = 0.100\ M$; e.g., 1/9 refers to $C_a = 0.01\ M$, $C_b = 0.09\ M$. Limiting lines *a* and *b* are for pure acid and base cases.

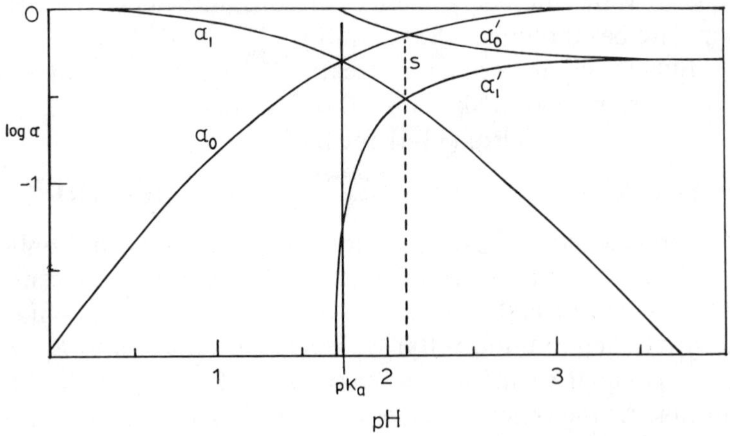

Figure 4-11. Expanded-scale log α diagram for 0.02 *M* HSO_4^-, 0.02 *M* SO_4^{2-} buffer showing both intersections at the same **pH**, *s*.

would be a mirror image set and would be equally useful in most circumstances. An example solution for a buffer requiring a quadratic equation is shown as Figure 4-11. This shows both intersections for a solution of 0.020 *M* $NaHSO_4$ and 0.020 *M* Na_2SO_4. The intersections of the two α_1 lines and of the two α_0 lines must agree in **pH** for any solution. Usually one is in a more easily read part of the diagram. Compare Figures 4-8–4-11. Generally, for buffers, the ratio method of Figures 4-1–4-4 is simpler to use. Placing the α curves of Figure 4-6 on the buffer diagram of Figure 4-10 with **pH** $=$ pK_a at the top will solve for the **pH** of all these solutions at the α_1 intersections.

3. Logarithmic Concentration Diagrams: **H** and **OH** Intersection Method

A method quite similar to the log α method just shown has recently become widely used in textbooks. It employs the same log α lines placed at the proper log C and pK_a position as just shown. This gives the log C for the species since we have by definition of the α fractions (C is the total X)

$$\mathbf{HX} = \alpha_1 C \qquad\qquad \mathbf{X} = \alpha_0 C$$

$$\log \mathbf{HX} = \log \alpha_1 + \log C, \qquad \log \mathbf{X} = \log \alpha_0 + \log C$$

Thus, when the α's are at their maximum, near 1, log α is zero and the log of the species is log C. Next in this method one adds the lines for log **H** and log **OH** for reference—these are already implied in the **pH** coordinate. **pH** problems are then solved by looking for points which satisfy the conditions of the problem. Usually these will be material or charge balance conditions. Thus, one does the same reasoning as in the log α method above, but in reverse order, which leads to complications in the cases for which the approximate equations (3-3)–(3-7) do not apply. The construction and use of these diagrams are illustrated in the following examples.

Example 3. Construct the log C diagram for all the species (except H_2O) present in the 0.100 *M* acetate system at any **pH**. Use it to deduce the **pH** of the pure acid and pure base solutions.

(a) Place the log α lines of Figure 4-6 upon the **pH**–log C coordinates with the system point, their intersection at **pH** $=$ pK_a and their top at log $C = -1$ for the 0.100 *M* system.

(b) Add the **H** and **OH** lines, 45° crossing at −7, 7. This is always true if pK_w is 14.00.

For this case let us use 0.1 *M* ionic strength again and put **pH** = pK_a = 4.56 and pK_w = 13.80. This moves the **OH** line 0.20 **pH** unit to the left and makes it intersect **H** at **pH** 6.90.

We now have log *C* lines for all four variable species of interest in this system (Figure 4-12). Figure 4-12 is called the log concentration diagram for the 0.1 *M* system. To solve for **pH**, look at the several intersections:

For pure 0.100 *M* acetic acid, charge balance requires

$$\mathbf{H} = \mathbf{OH} + [C_2H_3O_2^-]$$

In acidic solutions we see that the **OH** line is far below the others and **H** $\cong [C_2H_3O_2^-]$ is a good approximation, here. This equality occurs

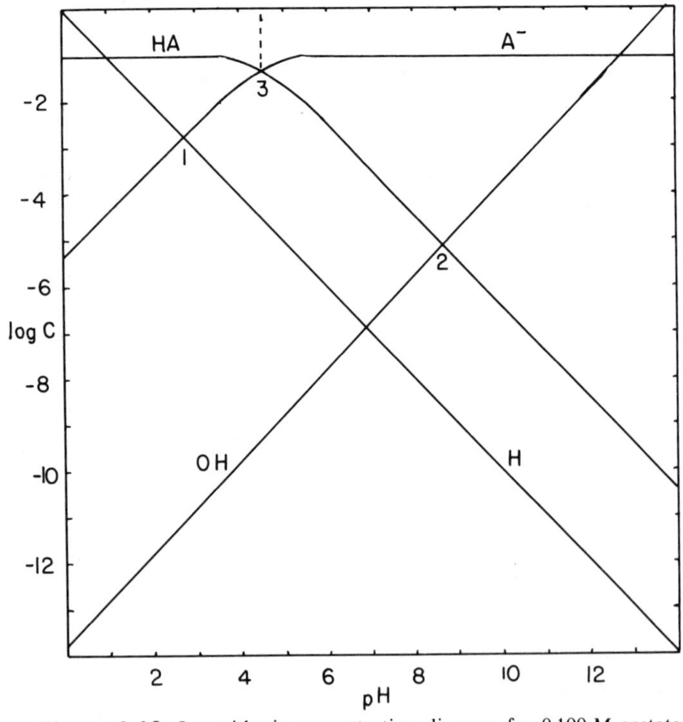

Figure 4-12. Logarithmic concentration diagram for 0.100 *M* acetate systems. Intersections: 1, pure acid; 2, pure base; 3, 1:1 buffer, the system point.

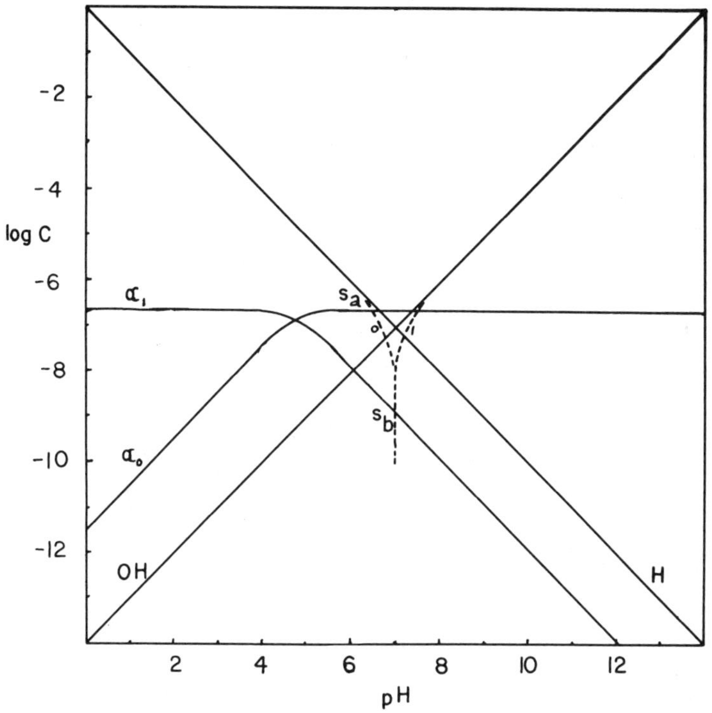

Figure 4-13. Comparison of log C and log α methods for 2×10^{-7} M acetate systems. The log α' lines are dashed where different from the **H** and **OH** lines of the log C method.

at intersection 1, at **pH** 2.8. For the 0.100 M conjugate base solution, from the derivation of equation (3-2) we have the proton balance condition

$$\mathbf{HX} = \mathbf{OH} - \mathbf{H}$$

If the **H** term is small, we have **HX** $\cong$ **OH**, which occurs at intersection 2, at **pH** 8.7. Here **HX** and **OH** are $10^{-5.1}$, which makes them compatible with the approximation made. For a 1:1 buffer, $C_a = C_b = 0.050\ M$, we are at the system point 3 at **pH** 4.56 if **H** and **OH** are much smaller than C_a and C_b, equation (3-2). This valid here, but becomes less so as the solution is diluted, as the α lines are lowered.

The advantage of these diagrams is that they can be easily constructed, mainly of straight lines and connecting curves, to show all the species in a solution at equilibrium. Let us show how these differ from the previous log α diagrams in a critical example.

Example 4. Construct and compare the log C and the log α diagrams for the acetate system at 2×10^{-7} M. Use $K°$ values. We slide the log α lines of Figure 4-12 down to the new log C, -6.70, and set $\mathbf{pH} = pK_a^\circ = 4.76$ at the system point intersection, and use 14.00 as the **pH** spread. This produces Figure 4-13, in which we have added in dashed lines the curved portion of the α' lines of Figure 4-7. That is the only difference between the two types of diagram. What is the difference in their uses? In this system we see that no terms are negligible in the charge and proton balance equations which we used above. Thus, the log C diagram only tells us that we must resort to algebraic methods to get a precise solution. Even a 1:1 buffer can be nowhere near the system point **pH** since the **H** would be much greater than the [HX] put into the solution. The log α' intersections with the log C lines do give the correct **pH** values for the pure conjugate acid and base solutions: pH 6.6 and pH 7.0. The log C diagram does show that acetate is effectively 2×10^{-7} M in both solutions, that is, $\alpha_0 \cong 1$. This value can be used in the numerical solutions if one does not proceed to the log α' method. In this case, where the full cubic equation form of (3-2) is required, only methods that take account of all terms will work. The log ratio and the log α' diagrams do this.

4. Summary

Three types of diagrams have been shown for describing monoprotic acid–base systems. These methods reinforce the conclusions of Chapter 3 about the importance of various terms in the complete equation (3-2). It is unlikely that one would plot these graphs from scratch to solve a problem. But, once available, they do help choose among approximate and more complete methods of solution.

Selected Reading

SILLEN, L. G., "Graphic Presentation of Equilibrium Data," in *Treatise on Analytical Chemistry*, ed. by I. M. Kolthoff and P. J. Elving, Interscience, New York, 1959, Part I, Vol. 1, Chapter 8. (Reprinted in paperback as *Chemical Equilibrium in Analytical Chemistry*.)

FREISER, H., and Q. FERNANDO, "Teaching Ionic Equilibrium: Use of Log-Chart Transparencies," *J. Chem. Ed.* **42**, 35–38 (1965).

LANGFORD, C. H., *BASIC Equilibrium Calculations*, Addison-Wesley, Reading, Massachusetts, 1973.

Problems

1. Use a straightedge to estimate the **pH** of the following solutions from Figure 4-3 or 4-4. Compare the results with the approximate equations and explain the differences. See the algebraic solutions in Chapter 3.
   a. 0.01 and 0.10 M trichloroacetic acid. $pK_a = 0.7$.
   b. 1:1 trichloroacetate buffers, $C_a = C_b = 0.10\ M$, and also 0.010 M.
   c. 0.010 and 0.10 M trichloroacetate alone, $C_a = 0$.
   d. A trichloroacetate buffer with $C_a = 0.20\ M$ and $C_b = 0.010\ M$.
   e. A trichloroacetate buffer with $C_a = 0.010\ M$ and $C_b = 0.10\ M$.

2. Use Figure 4-3 and a straightedge to estimate the **pH** of 0.10 M solutions of acids, bases, and buffers (1:1) of systems having pK_a 0, 2, 4, 6, 8, 10, and 12. Compare the results with those plotted in Figures 3-1 and 3-4.

3. Show numerically that the same shape α curves are obtained from equations (4-3) for the acetate system and the ammonium system using pK_a 4.7 and 9.3. Sketch these roughly on the same diagram and indicate the regions in which one expects to find (a) dilute acetic acid, (b) dilute NH_3 solutions, (c) dilute $NH_4C_2H_3O_2$ solutions, (d) a solution which has 0.010 M NH_3 and 0.10 M NH_4^+. Give reasons.

---

4. Show that equations (3-1) and (4-3) give the same result for the ratio α_1/α_0. Show what the difference is between the buffer ratio $R_b = C_a/C_b$ and the equilibrium ratio R_e. Which one is α_1/α_0?

5. Give a chemical interpretation of what is changing as one moves across the log α and log α' diagrams, Figures 4-6 and 4-7. How can these changes be brought about in practice?

---

6. In this chapter we used only α_0' or α_1' for specific cases, except in Figure 4-11. For pure acid and pure base cases, sketch all four lines, α_0' and α_1' for each, and compare the intersections for acetic acid with the ones in Figure 4-8.

7. Calculate enough points to sketch the log R_e, material balance, curves for pure acid and for pure base, each $2 \times 10^{-7}\ M$. Add the acetic acid R_e line and get the **pH** to compare with the other method in Figure 4-13, Example 4. Interpret your curves as to their limiting values.

8. Figure 3-3 shows that 1:1 buffers can change **pH** as they are diluted, yet Figure 4-1, which is also from a complete equilibrium calculation, shows that a **pH** uniquely fixes the ratio of conjugates. Explain why these are not inconsistent.

9. Make a tracing of Figure 4-10 and place upon it the log α lines of Figure 4-6 at pK_a 4.00. Read off the **pH** of the various buffers and compare with Figure 3-4.

10. How should one approach the problem of explaining monoprotic pH and buffering to those new to it? Which of the methods you have met do you find to give the clearest understanding of the factors in these equilibria? Is this necessarily the best approach for the novice?

5 Polyprotic Acid–Base Equilibria: Relations and Diagrams

In this chapter we show the general forms of multistep equilibria and their similarity for polyprotic acid–base and for polyligand metal ion–ligand equilibria. We consider only mononuclear cases, meaning that one central species accepts protons or ligands, as the case may be. Rarer polynuclear examples are H_2F_2 and Ag_2I^+.

1. Constants and Conventions

The first complication to deal with is the still-common treatment of acid–base equilibria in the "dissociation" direction using acidity constants:

$$HX + H_2O \rightleftharpoons H_3O^+ + X^-$$

$$K_a^\circ = (H_3O^+)(X^-)/(HX)$$

For polyprotic anions we shall use A subsequently. Other types of equilibria are now frequently, but not universally, treated as formations:

$$M + L \rightleftharpoons ML, \qquad K_1^\circ = (ML)/(M)(L)$$

where M is a metal ion (or any Lewis acid), and L is a ligand (or any Lewis base), charges not specified. For example, the first step of complexing in the silver(I)–ammonia system is

$$Ag^+ + NH_3 \rightleftharpoons Ag(NH_3)^+$$

$$K_1^\circ = (Ag(NH_3)^+)/(Ag^+)(NH_3)$$

There is a trend now to use formation directions for all cases, including acid–base.[1] However, acidity constants are so commonly used that we shall have to be able to shift between the systems. Let us look at examples of often used relations among the species in multistep equilibria.

(a) If a metal ion takes up to four ligands (charges not shown), we have

$$\begin{aligned}
&M + L \rightleftharpoons ML && K_1^\circ = \frac{(ML)}{(M)(L)} \\
&ML + L \rightleftharpoons ML_2 && K_2^\circ = \frac{(ML_2)}{(ML)(L)} \\
&ML_2 + L \rightleftharpoons ML_3 && K_3^\circ = \frac{(ML_3)}{(ML_2)(L)} \\
&ML_3 + L \rightleftharpoons ML_4 && K_4^\circ = \frac{(ML_4)}{(ML_3)(L)}
\end{aligned} \qquad (5\text{-}1a)$$

(b) If a base takes up to four protons (like the EDTA 4- ion)

$$\begin{aligned}
&H_4A \rightleftharpoons H^+ + H_3A^- && K_{a_1}^\circ = \frac{(H^+)(H_3A^-)}{(H_4A)} \\
&H_3A^- \rightleftharpoons H^+ + H_2A^{2-} && K_{a_2}^\circ = \frac{(H^+)(H_2A^{2-})}{(H_3A^-)} \\
&H_2A^{2-} \rightleftharpoons H^+ + HA^{3-} && K_{a}^\circ = \frac{(H^+)(HA^{3-})}{(H_2A^{2-})} \\
&HA^{3-} \rightleftharpoons H^+ + A^{4-} && K_{a_4}^\circ = \frac{(H^+)(A^{4-})}{(HA^{3-})}
\end{aligned} \qquad (5\text{-}1b)$$

Clearly, the one system is the reverse of the other, chemically, and the inverse, mathematically in the K's. In *Stability Constants of Metal Ion*

[1] See L. G. Sillen and A. E. Martell, *Stability Constants of Metal Ion Complexes*, Special Publications No. 17, 1964, and No. 25 (Supplement), 1971, The Chemical Society, London.

Complexes (see footnote 1) and many other international publications, the cases in (b) will be listed as in the (a) example, so that if one wants a traditional K_{a_1} for this acid one looks up K_4 for the species A^{4-}. That is,

$$A^{4-} + H^+ \rightleftharpoons HA^{3-}, \qquad K_1^\circ = \frac{(HA^{3-})}{(A^{4-})(H^+)} = (K_{a_4}^\circ)^{-1}$$

and so on to H_4A with $K_4 = (K_{a_1}^\circ)^{-1}$.

The two sets just given are in abbreviated chemical form. Water is critically needed in these reactions:

$$M(H_2O)_4 + L \rightleftharpoons M(H_2O)_3L + H_2O$$

$$H_4A + H_2O \rightleftharpoons H_3O^+ + H_3A^-$$

(As usual, we take water activity as *one* in dilute water solutions.) This does not change the mathematical dependence for these constants as we discussed in Chapter 2.

Overall constants, called betas, are sometimes useful. They are obtained by multiplying successive step constants:

$$\begin{aligned} \beta_1^\circ &= K_1^\circ \\ \beta_2^\circ &= K_1^\circ K_2^\circ = (ML_2)/(M)(L)^2 \\ \beta_3^\circ &= K_1^\circ K_2^\circ K_3^\circ = (ML_3)/(M)(L)^3 \\ \beta_4^\circ &= K_1^\circ K_2^\circ K_3^\circ K_4^\circ = (ML_4)/(M)(L)^4 \end{aligned} \tag{5-2}$$

Notice in set (5-2) how all the intermediate species cancel in these multiplications. While these betas are mathematically correct relations, they do not imply any corresponding single-step chemical reactions, as was formerly thought. Until it is proved otherwise, we may safely assume that all systems of the types shown in sets (5-1a) and (5-1b) are single-proton transfer, or ligand addition, step processes. To avoid confusion, we shall use β only for the formation direction. In protonic systems we shall write out products of acidity constants rather than trying to define a β in that direction.

2. Alpha Fractions and Diagrams

Because of the varied spacings of pK_a values, the polyprotic acid–bases do not have uniquely shaped α diagrams as do monoprotic systems. Let us see how to get the α expressions as functions of **H** for a triprotic case. Extension to other cases will be clear. Let the sum of all species of the system be C. For this system we have

$$C = [H_3A] + [H_2A] + [HA] + [A]$$

α_0, the fraction which is A (zero protons), is

$$\alpha_0 = \frac{[A]}{C} = \left[\frac{[H_3A]}{[A]} + \frac{[H_2A]}{[A]} + \frac{[HA]}{[A]} + 1 \right]^{-1}$$

The reciprocal form is used to give us simple ratios. These ratios are easily found in terms of **H** from the K_a expressions, set (5-1):

$$\frac{[H_3A]}{[A]} = \frac{H^3}{K_1K_2K_3}, \qquad \frac{[H_2A]}{[A]} = \frac{H^2}{K_2K_3}, \qquad \frac{[HA]}{[A]} = \frac{H}{K_3} \quad (5\text{-}3)$$

Substituting these into the α_0 expression above gives

$$\alpha_0 = \left[\frac{H^3}{K_1K_2K_3} + \frac{H^2}{K_2K_3} + \frac{H}{K_3} + 1 \right]^{-1}$$

The remaining α's can be obtained in the same way, or more simply from the α_0 already found:

$$\alpha_1 = \frac{[HA]}{C} = \frac{[A][HA]}{C[A]} = \alpha_0 \frac{[HA]}{[A]} = \alpha_0 \frac{H}{K_3}$$

$$\alpha_2 = \frac{[H_2A]}{C} = \frac{[A][H_2A]}{C[A]} = \alpha_0 \frac{[H_2A]}{[A]} = \alpha_0 \frac{H^2}{K_2K_3} \quad (5\text{-}4)$$

$$\alpha_3 = \frac{[H_3A]}{C} = \alpha_0 \frac{H^3}{K_1K_2K_3}$$

These rather complicated cubic **H** curves turn into symmetrical plots vs. p**H**.

For citric acid,

$$\overset{\displaystyle OH}{\underset{\displaystyle COOH}{HOOC-CH_2-\overset{|}{\underset{|}{C}}-CH_2-COOH}}$$

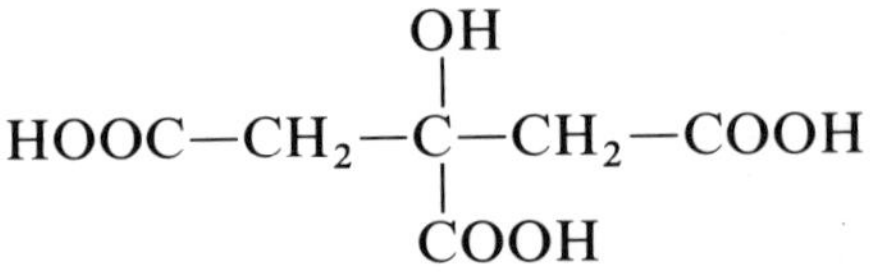

we have the following experimental constants at 25° and $I = 0.1\ M$:

$$pK_1 = 2.93, \qquad pK_2 = 4.36, \qquad pK_3 = 5.74$$

Table 5-1

pH	α_0	α_1	α_2	α_3
1	9.2×10^{-11}	5.0×10^{-6}	0.012	0.988
3	4.2×10^{-5}	0.023	0.528	0.448
5	0.128	0.708	0.162	0.002
7	0.947	0.052	1.2×10^{-4}	1.0×10^{-8}

There is a pK_4 at about 16 for the OH group which is not active in the low-**pH** solutions of interest here. Thus, this is effectively a triprotic acid. These are acidity constants.

Now one obtains the α values by substituting chosen **H** values in the expressions of equation set (5-4): for example, consider Table 5-1. Plotting a suitable number of such points vs. **pH** produces the diagram in Figure 5-1 for the variation of the four species with **pH**. It has some important features which are repeated in all the acid–base and M–L systems. The upper intersections occur at **pH** values equal to the pK_a values. Referring to set (5-1) shows why this must happen. When $\alpha_3 = \alpha_2$, then $[H_3A] = [H_2A]$, and $pK_1 = \mathbf{pH} = 2.93$. If one prepares a solution of NaH_2A aimed at the maximum of α_2, small and nearly equal amounts convert to H_3A and HA and one may expect the **pH** to be half-way between the pK_a values, $\frac{1}{2}(pK_{a_1} + pK_{a_2}) = 3.64$. However, this must be checked in the complete **H** equations. Even more uncertain from this diagram is the **pH** expected for solutions of H_3A or A alone. Other diagrams and algebraic methods will be introduced for these cases.

3. Log Ratio Diagrams

We introduced the monoprotic ratio methods in Chapter 4. We shall treat only the equilibrium condition ratio here. While the α's of Figure 5-1 show the variations of species in a most clear way, they are tedious to plot unless programmed for computer calculation and automatic plotter. The log ratio diagrams are composed entirely of straight lines and can be plotted in a few minutes. They contain all the information of the α diagrams, though in less clear display. We shall show how to derive the α diagrams from these log ratio plots. Let us do this for citric acid.

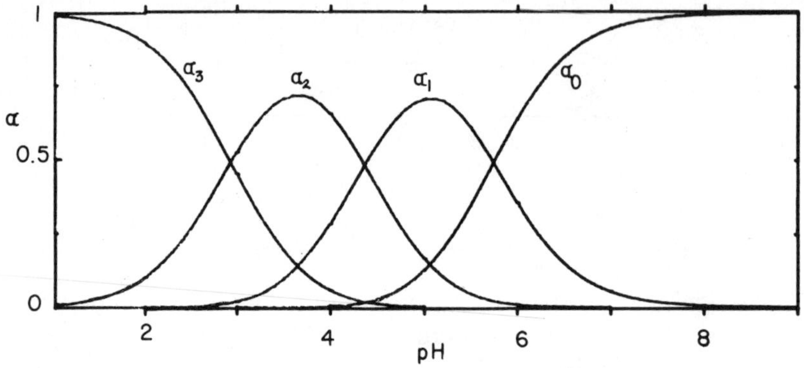

Figure 5-1. The α fraction curves for the four species in citrate systems at ionic strength 0.1 *M* and 25°C.

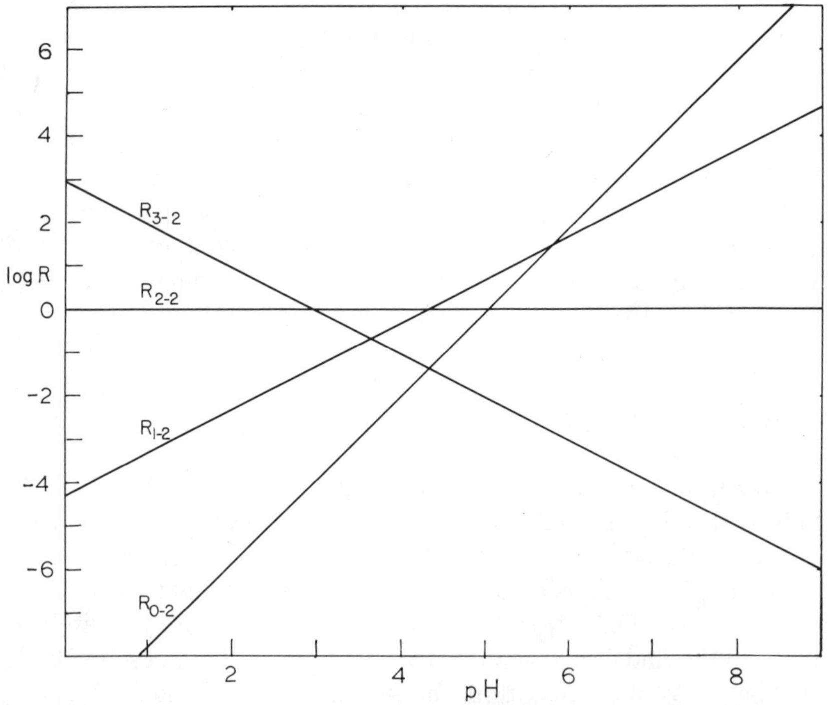

Figure 5-2. The log *R* curves for ratios of citrate species to H_2A^- (dihydrogencitrate ion) at ionic strength 0.1 *M*, 25°.

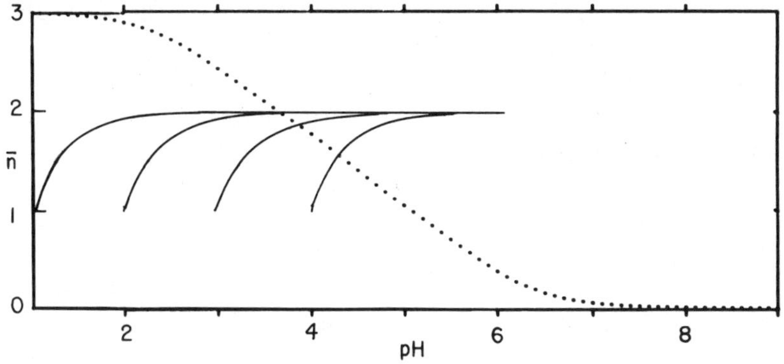

Figure 5-3. The equilibrium $\bar{n}$ (proton number; dotted) for the citrate system at ionic strength 0.1 *M*, 25°. The solid lines are four material balance $\bar{n}'$ lines for the solution of Example 1.

1. Get the species ratios and take their logs. [See equation sets (5-1) and (5-3).] The ratio between any species and a chosen reference species can always be expressed in terms of *K* values and **H** alone. We choose the central H_2A for convenience, for a more symmetrical diagram. We have

$$\begin{aligned} R_{3\text{-}2} &= [H_3A]/[H_2A] = \mathbf{H}/K_1, & \log R_{3\text{-}2} &= pK_1 - p\mathbf{H} \\ R_{2\text{-}2} &= [H_2A]/[H_2A] = 1, & \log R_{2\text{-}2} &= 0 \\ R_{1\text{-}2} &= [HA]/[H_2A] = K_2/\mathbf{H}, & \log R_{1\text{-}2} &= p\mathbf{H} - pK_2 \\ R_{0\text{-}2} &= [A]/[H_2A] = K_2K_3/\mathbf{H}^2 & \log R_{0\text{-}2} &= 2p\mathbf{H} - pK_2 - pK_3 \end{aligned} \tag{5-5}$$

2. The diagram. Place the reference horizontal for $\log R_{2\text{-}2}$ at the center (Figure 5-2). Next, add $\log R_{3\text{-}2}$, which is a line of slope -1 and intercept pK_1 at $p\mathbf{H} = 0$. It must also pass through $p\mathbf{H} = pK_1$ at $\log R_{3\text{-}2} = 0$. For the others we have, respectively, for slope and intercept at $p\mathbf{H} = 0$

$$\log R_{1\text{-}2}: \quad +1 \quad -pK_2$$

$$\log R_{0\text{-}2}: \quad +2 \quad -p(K_2K_3)$$

Comparison of this diagram with Figure 5-1 shows that the intersections occur at the p*K* values, as they should. (When two ratios are equal, their numerators are equal, and thus the α values are equal.) The other intersections also agree with the algebra: e.g., $\alpha_3 = \alpha_0$

when $R_{3\text{-}2} = R_{0\text{-}2}$, which occurs, according to set (5-5), at **pH** = $\frac{1}{3}$p$(K_1K_2K_3)$, or **pH** 4.34.

Both Figures 5-1 and 5-2 show the relative amounts of species at any **pH**. This is helpful in choosing approximate numerical methods. For example, from **pH** 1 to 3, H_3A (α_3) and H_2A (α_2) are the major species. $R_{1\text{-}2}$ is two log units below $R_{3\text{-}2}$ and $R_{2\text{-}2}$, so α_1 is under 1% of α_2 and α_3. We may correctly conclude that the K_1 expression alone describes the major citrate **pH** effects in this **pH** range. However, near **pH** 4, H_3A, H_2A, and HA are all significant, so that we need both K_1 and K_2 to treat the major species in equilibrium.

The citric acid α diagram could be constructed from the ratios read from Figure 5-2 as follows. At **pH** from 0 to 2, the α_3 line must be above 0.9 since $R_{3\text{-}2}$ is more than one log unit above any other R. Next we choose simple ratios to examine. At low **pH** where α_0 and α_1 are still insignificant on the α plot, log $R_{3\text{-}2} = 1$ (at **pH** 1.93) tells us that $R_{3\text{-}2} = 10/1$, so that $\alpha_3 = 10/11$ and $\alpha_2 = 1/11$, or 0.91 and 0.09. Then α_3 and α_2 must cross at **pH** = pK_1 = 2.93.

$R_{3\text{-}2}$ and $R_{1\text{-}2}$ cross at **pH** 3.64 and log R – 0.72 ($R = 0.19$) in Figure 5-2. This tells us that $\alpha_3/\alpha_2 = \alpha_1/\alpha_2 = 0.19$. Since $R_{0\text{-}2}$ is small here, we may approximate (total fractions = 1)

$$\alpha_3 + \alpha_2 + \alpha_1 = 1$$

$$0.19\alpha_2 + \alpha_2 + 0.19\alpha_2 = 1$$

$$\alpha_2 = 0.72 \qquad \text{and} \qquad \alpha_1 = \alpha_3 = 0.14$$

Continuing in this way, we can construct an approximate α diagram suitable for many purposes.

Another useful approximation in this construction is that the maxima of α_2 and α_1 occur over the crossings of the adjacent species. [This can be proved by differentiation of the α expressions, set (5-4).] If the maximum of α_2 occurs over the crossing of α_3 and α_1, we can set equal the expressions for $R_{3\text{-}2}$ and $R_{1\text{-}2}$ [set (5-5)] to find simply that $\mathbf{H} = (K_1K_2)^{1/2}$. Next we substitute this **H** in the α_2 expression in set (5-4), omitting the final term, which is the small contribution from the a_0. This gives

$$\alpha_{2\max} = (\mathbf{H}/K_1 + 1 + K_2/\mathbf{H})^{-1} = [1 + 2(K_2/K_1)^{1/2}]^{-1}$$

Putting in the K values for citric acid, we find $\alpha_{2\max} = 0.72$. Similarly, $\alpha_{1\max}$ occurs approximately at **pH** 5.05 at

$$\alpha_{1\max} = [1 + 2(K_3/K_2)^{1/2}]^{-1} = 0.71$$

4. Graphical Solutions to Polyprotic pH Problems

If we could know the equilibrium fractions α of the species present, it would be a simple matter to find the **pH** of solutions. However, we usually know only the initial materials, the analytical concentrations. Algebraic methods of approximation will be given in the next chapter. The graphical methods here help in understanding these equilibria and in evaluating the approximate equations to be found later.

We follow a similar course to that in Chapter 4, where the intersection of equilibrium and material balance fraction functions solves the problem. Here, the fraction of combined protons is needed. Take the sum of all available **H** acidity (using a triprotic example)

$$C_{\mathrm{H}} = [\mathrm{HA}] + 2[\mathrm{H_2A}] + 3[\mathrm{H_3A}] + \mathbf{D} \qquad (5\text{-}6)$$

For example, a 0.1 M H_3A solution should have $C_{\mathrm{H}} = 0.3\ M$ but distributed in an unknown proportion among the terms on the right-hand side of equation (5-6). We can rearrange this to get two expressions for the bound proton concentration:

$$[\text{bound acidity}] = C_{\mathrm{H}} - \mathbf{D} = [\mathrm{HA}] + 2[\mathrm{H_2A}] + 3[\mathrm{H_3A}]$$

Now, $C_{\mathrm{H}} - \mathbf{D}$ is just the material balance statement that the bound acidity is the total acidity minus the unbound acid **D**. The other term is the sum of the equilibrium species binding protons. Next we obtain our ratio of bound acidity to the total A species C. This is the average number of protons per A, called $\bar{n}$, a very useful experimental function. Divide by C:

$$\frac{[\mathrm{HA}] + 2[\mathrm{H_2A}] + 3[\mathrm{H_3A}]}{C} = \frac{C_{\mathrm{H}} - \mathbf{D}}{C}$$

We call the first $\bar{n}$, and the second $\bar{n}'$, having the same significance as in Chapter 4, for equilibrium condition and for material balance condition, respectively. From the definitions of the α fractions in set of equations (5-4), we can write these as

$$\alpha_1 + 2\alpha_2 + 3\alpha_3 = \bar{n}, \qquad (C_{\mathrm{H}} - \mathbf{D})/C = \bar{n}' \qquad (5\text{-}7)$$

The complete expressions for α's and **D** in terms of **H** [set (5-4)] would give us a correct fifth degree equation in **H**, putting $\bar{n} = \bar{n}'$. [In general, the complete equation for an n-protic system is of $(n + 2)$th degree in

H.] Numerical solutions are traditionally accomplished by approximations. How to make valid approximations should become clear after study of relevant α diagrams and graphical methods. Clearly we have the same situation as in Chapter 4. We can plot the two $\bar{n}$ functions to find their simultaneous solution instead of solving the fifth degree equation. Once the system (K_a's) are chosen, $\bar{n}$ vs. **pH** is defined. Once C and C_H (the materials put into solution) are chosen, the $\bar{n}'$ vs. **pH** curve is defined. If α diagrams, or values, or computer programs are already available for the substance in question, this may be the best way to solve a **pH** problem. The method to follow illustrates all the variety of polyprotic solution situations that can arise and should clarify the valid approach to understanding them. We again have a function of invariant shape in $\bar{n}'$, but not in $\bar{n}$.

Example 1. Examine the **pH** of NaH_2citrate solutions at analytical concentrations from 0.10 M to 10^{-4} M. Assume 0.1 M ionic strength and visualize the initial species placed in solution as Na^+ and H_2 citrate$^-$ ions.

Figures 5-1 and 5-2 allow us to decide that, at most, only 72% of the citrates can remain as this H_2A^- ion. If we are near this maximum, as shown above, the **pH** is near 3.64 [$(K_1K_2)^{1/2}$] and α_0 is small. As in the approximation equation (3-5), this shows no relation to concentrations, and so cannot be complete, since all solutions approach neutrality as $C \longrightarrow 0$. Let us see what $\bar{n}'$ and $\bar{n}$ in the complete expression, equation (5-7), add to this.

The equilibrium $\bar{n}$ curve for this citrate system is shown as the dotted plot in Figure 5-3. [This could be obtained from α values read from Figure 5-1 and used in the left expression of equation (5-7).] It shows the average number of protons per citrate, starting from three in acid solution, and falling toward zero in basic solution. If the approximate answer above is valid, we are at $\bar{n} = 2.00$. This follows if $\alpha_3 = \alpha_1$. Calculate $\bar{n}'$ for our solutions with $C_H = 2C$, and $C = C$ in equation (5-7):

$$\bar{n}' = C_H/C - \mathbf{D}/C$$

$$\bar{n}' = 2 - 10\mathbf{D} \qquad \text{for } 0.10\ M \text{ solution}$$

$$\bar{n}' = 2 - 100\mathbf{D} \qquad \text{for } 0.010\ M \text{ solution}$$

$$\bar{n}' = 2 - 1000\mathbf{D} \qquad \text{for } 0.0010\ M \text{ solution}$$

Table 5-2

pH	pOH	$\bar{n}$		
		C = 0.10 *M*	0.010 *M*	0.0010 *M*
1.0	—	1.0	—	—
1.3	—	1.5	—	—
1.7	—	1.8	0	—
2.0	—	1.9	1.0	—
2.3	—	1.95	1.5	—
2.7	—	1.98	1.8	0
3.0	—	1.99	1.99	1.0
4.0	—	2	1.99	1.9
6.9	—	2	2	2
9.8	4.0	2	2.01	2.1
10.8	3.0	2.01	2.1	3.0
11.8	2.0	2.1	3.0	12.0
12.8	1.0	3.0	12.0	102.0

D is effectively **H** in acid solutions and **−OH** in basic solutions. We must use **D** = **H** − **OH** near neutrality. The **pH** values in Table 5-2 demonstrate the invariant shape of this function. The plot of such a curve is shown in Figure 5-4 for the case of $C = 0.010\,M$ NaH_2citrate, $C_H/C = 2$. The basic portion can be understood to arise from possible basic behavior; the ion accepts protons from water, producing OH^-. This is an inverted mirror image of the acid half. We centered around **pH** 6.9, for $pK_w = 13.80$ for 0.1 *M* ionic strength. The invariant portion shown boxed in Figure 5-4 needs only to be placed upon the equilibrium $\bar{n}$ diagram so that its horizontal lies along $\bar{n} = C_H/C$, and with the cross mark at $\bar{n} = (C_H/C - 1)$, at **pH** = $-\log C$. In basic cases, the box can be inverted and rotated 180° and be placed at corresponding positions: cross at $(C_H/C + 1)$ at **pOH** − $\log C$. Thus we can shift the $\bar{n}'$ curve along the **pH** axis to read off the intersection values for any chosen C in our problem. This is shown in Figure 5-3, where relevant portions of this $\bar{n}'$ are shown for the 0.10 to 10^{-4} *M* H_2citrate cases. The intersections occur at **pH** values 3.64, 3.66, 3.83, and 4.34 for these four concentrations a factor of 0.1 apart. Thus,

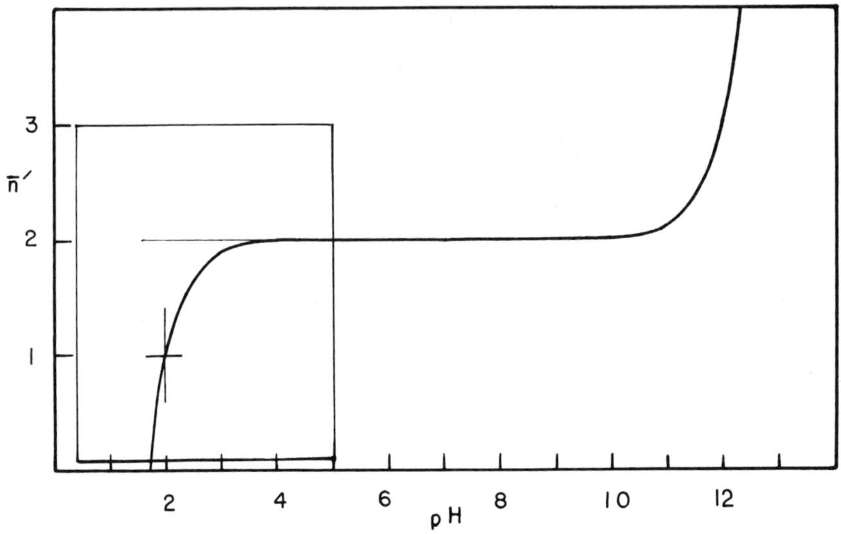

Figure 5-4. The material balance $\bar{n}'$ plot for the case $C_H = 0.020\ M$, $C = 0.010\ M$. The boxed portion is invariant in shape for C above $10^{-5}\ M$.

we find that the approximate ampholyte equation, **pH** $= (K_1K_2)^{1/2}$, holds above 0.01 M while the full $\bar{n}$ treatment correctly predicts an approach toward neutrality with dilution.

It is now a simple matter to use the same method to solve any problem in the citrate system, by moving the $\bar{n}'$ fragment to the proper position in Figure 5-3. For example:

1. A 1:1 buffer 0.0050 M in each of citric acid and NaH_2citrate gives us $C_H = 0.0250\ M$ and $C = 0.010\ M$. We have $\bar{n}' = 2.5 - 100\mathbf{D}$. Placing the horizontal of $\bar{n}'$ (Figure 5-4) on 2.5 and the crossed point at $\bar{n} = 1.5$ and **pH** 2.00 allows us to read off the intersection at **pH** 3.05. The approximate buffer equation (3-5) predicts **pH** $= pK_1 =$ 3.13.

2. Pure 0.010 M citric acid has $C_H = 0.030\ M$ and $C = 0.010\ M$. We have $\bar{n}' = 3 - 100\mathbf{D}$. Placing the horizontal of $\bar{n}'$ at 3 and the crossed point at $\bar{n} = 2.00$ and **pH** 2.00 gives the intersection at **pH** 2.51.

This method becomes impractical for pure trisodium citrate and concentrated citric acid solutions, where both the $\bar{n}$ and the $\bar{n}'$ curves are nearly horizontal, below pH **2** and above **pH** 8. But in these regions approximations treating the species as single monoprotic donors or acceptors work. Figure 5-1 shows that K_1 governs below pH 2 (α_1 and α_0 are negligible) and that K_3 governs above pH 6 (α_2 and α_3 are negligible).

Figure 5-5 shows the contrasting α diagram for the phosphate system with pK_a values 1.95, 6.80, and 11.67 at 25° and $I = 0.1\ M$. This looks like three separate monoprotic cases (Figure 4-5). Indeed,

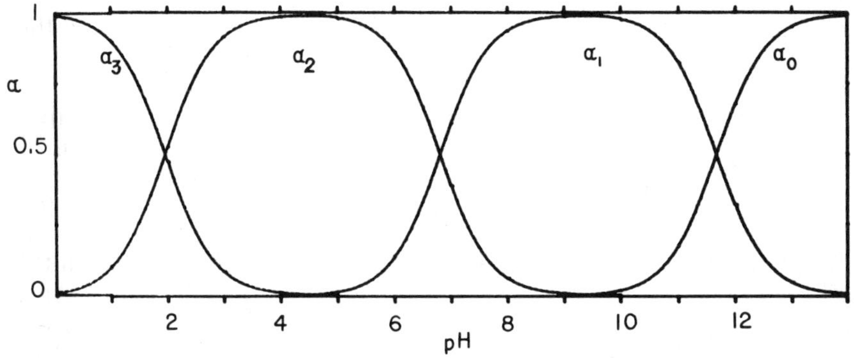

Figure 5-5. The α fraction curves for the phosphate system at ionic strength 0.1 M at 25°C.

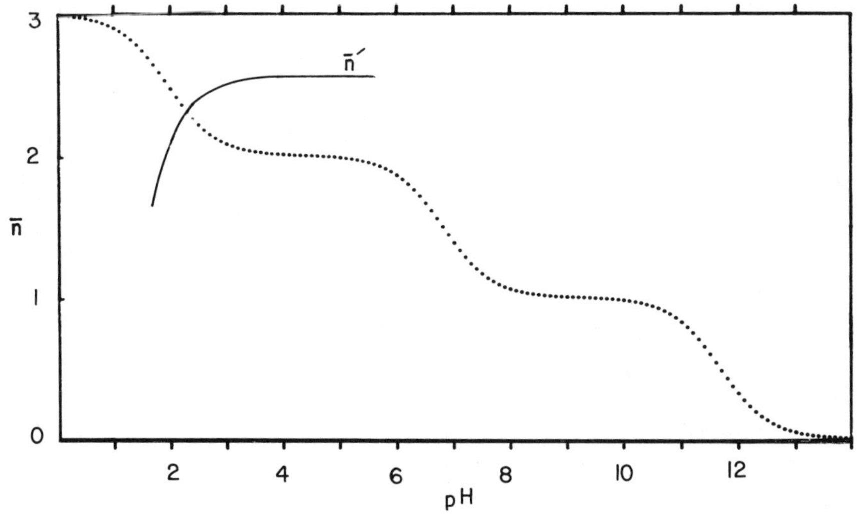

Figure 5-6. The equilibrium $\bar{n}$ curve (dotted) for the phosphate system at ionic strength 0.1 M at 25°. The solid line is $\bar{n}'$ for the solution of Example 2.

most phosphate buffers can be treated accurately using simple ratio approximations. The ampholytes, near the maxima of α_2 and α_1, will require two constants for accurate results. The phosphate $\bar{n}$ curve (Figure 5-6) again emphasizes the difference from the citrates (Figure 5-3).

The four possible situations which occur in polyprotic systems may be simply described in terms of the ratio of available acidity to the total anion, C_H/C. Methods of approaching **pH** problems for all these types are given in this chapter and in Chapter 6.

Pure acids:	$C_H/C = n$, the maximum number of protons the anion will accept.
Pure bases:	$C_H = 0$, the anion alone.
Ampholytes:	$C_H/C = 1, 2, \cdots$ up to $n - 1$, integral values only.
Buffers:	$C_H/C =$ nonintegral values between zero and n.

As in monoprotic cases in Chapter 3, we stress the necessity of allowing possible neutralizations to go to completion before computing C_H. For example, mixing 6 mmol of H_3PO_4, 1 mmol Na_3PO_4, and 3 mmol NaOH in 100 ml yields a C_H of 0.15 and $C = 0.07$, which give a maximum possible $\bar{n}$ of 2.14, a buffer solution in the region governed largely by K_1. When the K values are widely separated, this ratio 2.14 tells us that H_3A and H_2A^- must be the major species. It should be helpful to focus on this ratio C_H/C in each example as an aid in orienting one to the proper **pH** region in these systems.

Example 2. Find the **pH** of a buffer made to be 0.030 M in H_3PO_4 and 0.020 M in NaH_2PO_4. Here we have $C_H = 0.130\ M$ and $C = 0.050\ M$. In equation (5-7) this gives us

$$\bar{n}' = 2.60 - 20\mathbf{D}$$

This is plotted over the equilibrium $\bar{n}$ in Figure 5-6. The intersection is at **pH** 2.05, and α_3 and α_2 are 0.44 and 0.56. The approximate buffer ratio might lead us to expect these to be 0.6 and 0.4 at **pH** 1.77, using the K_1 expression. Thus, the $\bar{n}$ method has solved the quadratic for us in this case.

Mixtures of two systems are most difficult to treat by any means other than the graphical, or computer, methods based upon the $\bar{n}$ function. For a final example in this section, we look at such a mixture published in handbooks as McIlvaine's buffers.

Example 3. The published **pH** 5 mixture requires 9.70 ml of 0.100 *M* citric acid and 10.30 ml of 0.200 *M* Na_2HPO_4. Let us dilute this to a total volume of 60.0 ml to obtain an ionic strength about 0.1 so that our $\bar{n}$ diagrams are valid. Taking into account these dilutions, the analytical concentrations taken are

total citrates	$C_c = 0.0162\ M$
total phosphates	$C_p = 0.0343\ M$
total acidity	$C_H = 3C_c + C_p = 0.0829\ M$

After equilibrium is reached, the acid balance is

$$C_H = \bar{n}_p C_p + \bar{n}_c C_c + \mathbf{D}$$

(At **pH** near 5, **D** = **H**.) Inserting the values given, we obtain

$$C_H = 0.0829 = \bar{n}_p(0.0343) + \bar{n}_c(0.0162) + \mathbf{H}$$

Now, we have only to read off $\bar{n}$ values for chosen **H** on the curves of Figures 5-3 and 5-6 until satisfactory closeness in balancing this equation is achieved. For example, at **pH** 5.00 we get

$$C_H = 1.98(0.0343) + 1.03(0.0162) + 10^{-5} = 0.846$$

At **pH** 5.10 we get $C_H = 0.830$. By plotting, or simple interpolation, we soon find that 5.11 satisfies the C_H equation. The ionic strength at these proportions turns out to be 0.08 *M*. Taking the f_+ for hydronium as 0.83, we get 5.19 as the expected pH meter reading for this diluted buffer. Note that, in agreement with the diagrams for these acids, the $\bar{n}$ values show us that citrate is mainly HA^{2-} and phosphate is H_2A^- in the resulting solution.

5. Logarithmic Fraction and Concentration Diagrams

Plots of the logarithms of α values for the citric and phosphoric acid systems above are shown in Figures 5-7 and 5-8. Since, by definition, $\alpha_n C = [H_nA]$, we can write

$$\log [H_nA] = \log \alpha_n + \log C$$

Thus, we can obtain the log C_n line for any species H_nA by moving the log α diagram log C units as shown in Figure 5-9, in which Figure 5-8 has been lowered two units to produce the log C diagram for the 0.010 M phosphate system. As with the log C diagrams for monoprotic systems in Chapter 4, we can locate approximate **pH** regions for various solutions of given analytical concentrations and make decisions about negligible species to help in calculating refined results. The **H** and **OH** lines can be added as before to assist. Let us examine two examples at extremes at which the n method above is less practical.

Example 4. Find the **pH** of pure 0.010 M H_3PO_4 solution. Assume total ionic strength at 0.1 M and that **D** = **H** in these acid solutions.

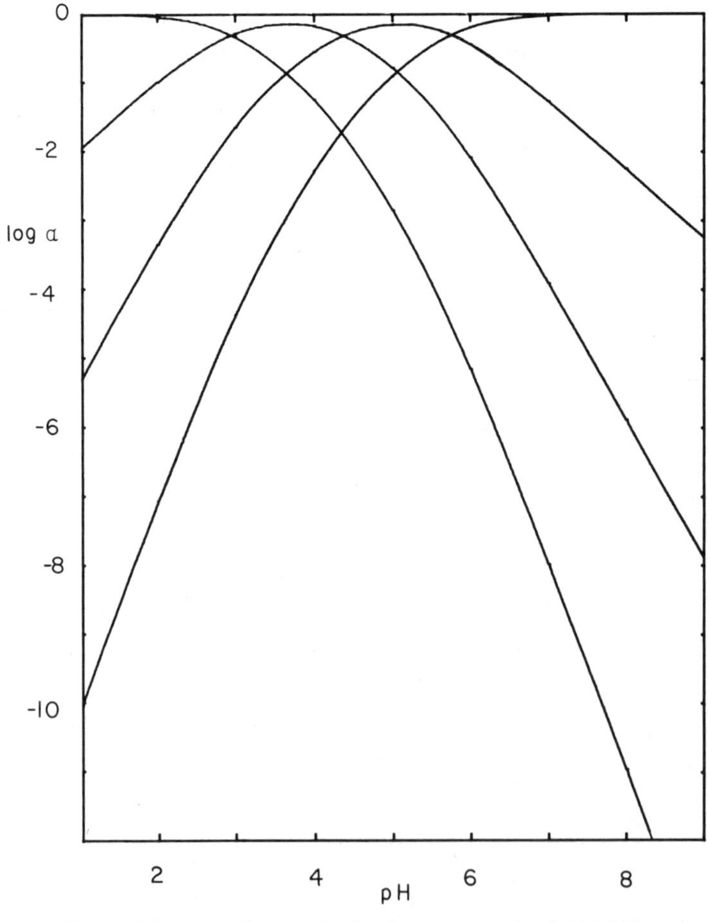

Figure 5-7. Log α diagram for the citrate system, $I = 0.1\ M$, 25°.

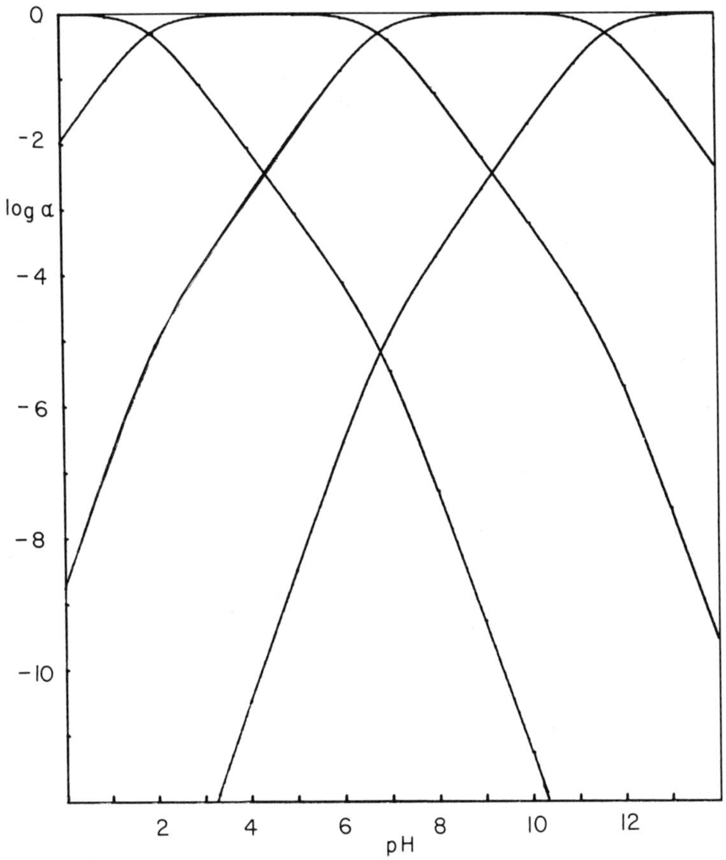

Figure 5-8. Log α diagram for the phosphate system at $I = 0.1\ M$ at 25°.

Charge balance requires

$$\mathbf{H} = \mathbf{OH} + [H_2PO_4^-] + 2[HPO_4^{2-}] + 3[PO_4^{3-}]$$

But the diagram (Figure 5-9) shows that α_1, α_0 and **OH** are all very small where α_3 (H_3PO_4) is large, so that we have

$$\mathbf{H} \cong [H_2PO_4^-]$$

This crossing occurs at **pH** 2.20, the antilog of which gives

$$\mathbf{H} = [H_2PO_4^-] = 0.0063 \qquad \text{and} \qquad [H_3PO_4] = 0.0037\ M$$

These check in the K_1 expression.

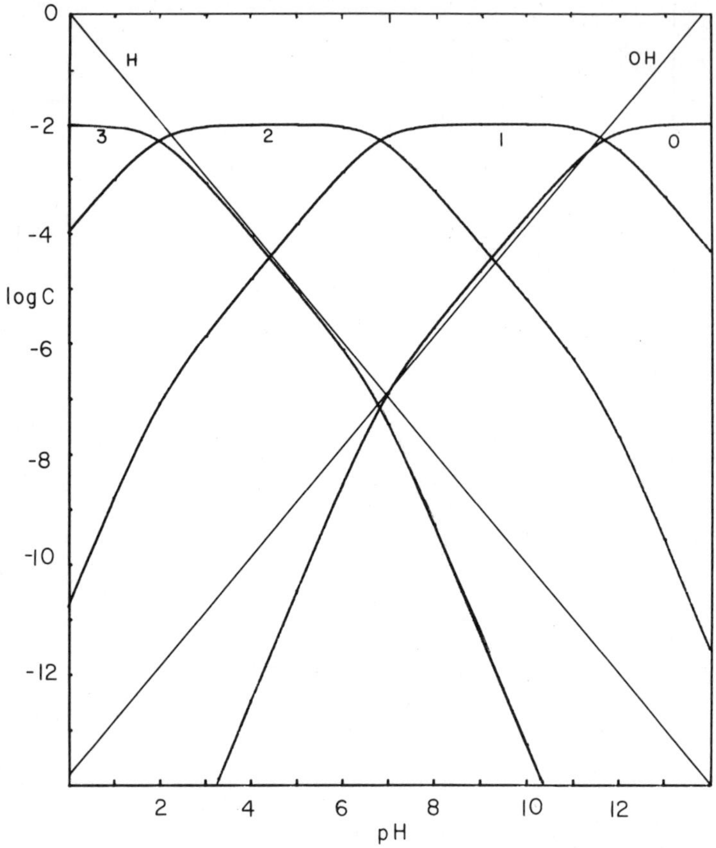

Figure 5-9. The logarithmic concentration diagram for the system 0.010 *M* phosphates at $I = 0.1$ *M* at 25°. The numbers refer to the concentration lines for H_3PO_4, $H_2PO_4^-$, HPO_4^{2-}, and PO_4^{3-}. Line 3 is for H_3PO_4, etc.

Example 5. Find the **pH** of pure 0.010 *M* Na_3PO_4 solution at 0.1 *M* ionic strength at 25°. The combined material and equilibrium condition balances of equation (5-7) help here. $C_H = 0$ in this case, and $\mathbf{D} = -\mathbf{OH}$ at high **pH**. We have

$$-\mathbf{D}/C = \alpha_1 + 2\alpha_2 + 3\alpha_3 \cong \alpha_1$$

since α_2 and α_3 are very small when α_0 and α_1 are large. Thus, we need to find the intersection of log **OH** and log $[HPO_4^{2-}]$. We have

$-\mathbf{D} = \mathbf{OH} = \alpha_1 C$. This occurs at **pH** 11.55, which gives the result (using $pK_w = 13.80$)

$$\mathbf{OH} = [HPO_4^{2-}] = 10^{-2.25} = 0.0056 \qquad \text{and} \qquad [PO_4^{3-}] = 0.0044$$

These check in the K_3 expression.

Various phosphate buffers can be treated in a similar way. In general, use of log C diagrams for polyprotic systems requires imaginative decisions about the region, major species, and what relations to look for on the diagram. No one set of steps guarantees quick, correct results, as we can get with the $\bar{n}$ method. With any method, the results should be checked against diagrams, K expressions, and common sense.

Ionic Strength Effects. The computer plots of α and $\bar{n}$ for the citric acid system at ionic strength zero and 0.10 M are shown in Figures 5-10 and 5-11. These clearly demonstrate that ionic strength affects the more highly charged ionic equilibria more than the lower charged cases. Each successive K_a becomes more acidic, more ion-forming, in going to higher ionic strength in this low concentration region which follows Debye–Hückel behavior to fair approximation. The $\bar{n}$ value is correspondingly less at each **pH** value. In the case of the citrate buffer mentioned above, 0.0050 M in both citric acid and NaH_2 citrate, the **pH** at I 0.1 M was 3.05. If we use the zero-I curve for $\bar{n}$, the result is 3.22.

6. Microscopic versus Macroscopic Equilibrium Constants

In general, we deal with the apparent macroscopic, step equilibrium constants. When there are different base positions of similar proton affinity, the distribution among different species of the same charge may be an important consideration. For example, consider cysteine:

$$HS{-}CH_2{-}CHNH_3^+{-}COO^-$$

The three macroscopic acidity constants determined by standard pH methods are $pK_1 \cong 2.0$, $pK_2 = 8.33$, and $pK_3 = 10.50$ at 20° and I = 0.1 M ($NaClO_4$). There is no ambiguity in K_1 for the carboxyl proton in the protonated molecule in very acid solution (pH 2).

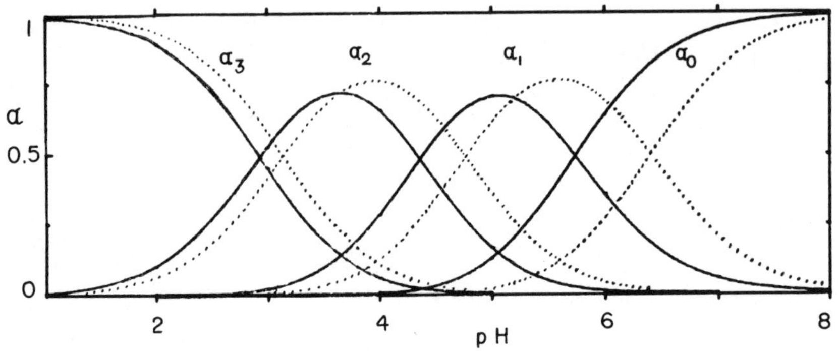

Figure 5-10. Two sets of α curves showing ionic strength effects in the citrate system. Solid lines are for $I = 0.1\ M$, and dotted lines are for the zero-ionic-strength limit, 25°.

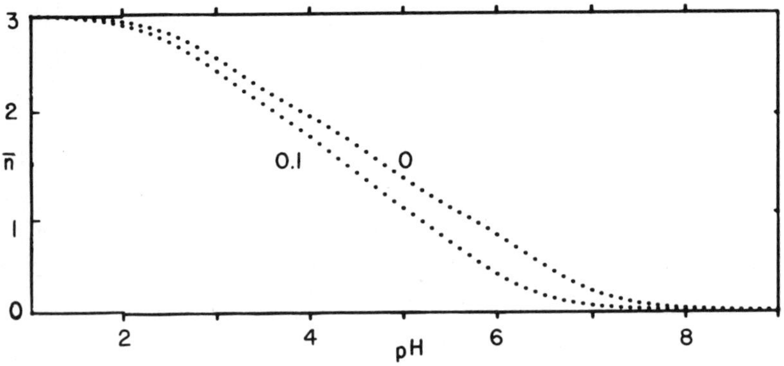

Figure 5-11. Proton number curves for the citrate system at 0.1 and 0 ionic strengths.

However, in basic solution, four species may exist in appreciable fractions together because the $-NH_2$ and $-SH$ groups are of similar proton affinity:

$$\begin{array}{ccc} HS-CH_2-CHNH_3^+-COO^- & \overset{a}{\rightleftharpoons} & {}^-S-CH_2-CHNH_3^+-COO^- \\ {\scriptstyle b}\upharpoonleft\downharpoonright & & \upharpoonleft\downharpoonright{\scriptstyle c} \\ HS-CH_2-CHNH_2-COO^- & \overset{d}{\rightleftharpoons} & {}^-S-CH_2-CHNH_2-COO^- \end{array}$$

The macro-constants K_2 and K_3 do not distinguish between the two -1 anions, but use only their sum. Setting up the usual equilibrium expressions for the micro-constants, one can then show that $k_a/k_b = k_d/k_c$ and that $K_2 = k_a + k_b$ and $K_3^{-1} = k_c^{-1} + k_d^{-1}$.

Experimental determination of the values of micro-constants is difficult and few are available. For cysteine, this has been done[2] by comparing spectra at several pH values with those for similar molecules with fewer functional groups (—SH, $-NH_2$, —COOH). The results deduced for 23° (I variable but below 0.1 M) were $pk_a = 8.53$, $pk_b = 8.86$, $pk_c = 10.36$, and $pk_d = 10.03$. These lead to the macro-values $pK_2 = 8.36$ and $pK_3 = 10.53$, in reasonable agreement with those determined from pH methods mentioned above. The α diagram for such a system varies from those we have drawn by having the α_2 portion divided between the two parts, in this case about $\frac{2}{3}$ with the proton on $-NH_3^+$ and $\frac{1}{3}$ with it on —SH. The α diagram is shown in the reference cited in footnote 2.

A detailed treatment of this topic is given by King.[3]

7. Summary

Species fractions, α's, and ratios between the members of any pair of species in equilibrium in mononuclear polyprotic systems have been shown to depend only upon **H** and the K values. Uses of plots of these functions have been illustrated in problem solving. The $\bar{n}$ function related to α's is useful because it can be determined experimentally from analytical concentrations and a **pH** measurement. It also provides a fundamentally complete and rigorous approach to **pH** calculations.

The **pH** in all diagrams refers to the molarity function because of the requirement for charge and material balancing in deriving some results. Conversion to activity can be accomplished after finding answers in molarity hydronium concentration.

Selected Reading

BUTLER, JAMES N., *Ionic Equilibria*, Addison-Wesley, Reading, Massachusetts, 1964, Chapter 7.

DE LEVIE, R., Ratio diagrams, *J. Chem. Ed.* **47**, 187 (1970).

---

[2] R. E. BENESCH and R. BENESCH, *J. Am. Chem. Soc.*, **77**, 5877 (1955).

[3] E. J. KING, *Acid-Base Equilibrium*, Macmillan, New York, 1965, Chapter 9.

FREISER, HENRY and Q. FERNANDO, *Ionic Equilibrium in Analytical Chemistry*, Wiley, New York, 1963, Chapters 4–7.
GUENTHER, W. B., *Quantitative Chemistry*, Addison-Wesley, Reading, Massachusetts, 1968, Chapters 5 and 8.
RICCI, JOHN E., *Hydrogen Ion Concentration*, Princeton University Press, Princeton, New Jersey, 1952.

Problems

1. From the log R diagram (Figure 5-2), estimate the **pH** at which α_3 should fall to about 0.01 and check this on Figure 5-1. Repeat this at both ends of α_2 and α_1 to clarify the relation of these diagrams.

2. Use Figures 5-1 and 5-5 to estimate the **pH** regions in which you expect the following solutions to fall. Which can be rather precisely located and which not? Use the K expressions to show why this is true at the expected **pH** values. For citrates and phosphates: H_3A, KH_2A, K_2HA, K_3A.

3. Construct a log R diagram for oxalic acid systems at 0.1 M ionic strength. Estimate the **pH** of a buffer which is 0.010 M in oxalic acid and 0.100 M in $NaHC_2O_4$. Check your result by looking into the charge and material balance of the equilibrium species. Take the published conditional pK_a values as 1.13 and 3.85 at 25°.

4. Sketch approximate α and $\bar{n}$ diagrams for oxalic acid systems at $I = 0.1$ (constants in previous problem). Deduce the **pH** of 0.010 M solutions of (a) oxalic acid, (b) $NaHC_2O_4$, (c) equimolar $H_2C_2O_4$ and $NaHC_2O_4$ both 0.010 M.

---

5. Deduce the general form of the titration curves for 0.100 M solutions of citric and of phosphoric acid by 1.000 M NaOH by examination of their α and $\bar{n}$ curves in this chapter. Sketch the curves: **pH** vs. ml NaOH added for 100 ml of acid.

6. Use Figure 5-11 and the $\bar{n}'$ intersection method to estimate the change in **pH** when 0.1 mol per liter of NaCl is dissolved in 0.001 M citric acid. Assume that this is near zero ionic strength to start. Repeat for Na_3citrate solution.

---

7. Differentiate the α_2 and α_1 expressions in equation (5-4) with respect to **H**. Then put in the K values for citric acid and find the **H** at the maxima by setting the derivatives equal to zero.

8. Construct a log ratio diagram for the $I = 0.1$ M phosphate system. Estimate the **pH** of 0.010 M Na_2HPO_4 from it. Test the result on the $\bar{n}$ diagram (Figure 5-6) and on the log α diagram (Figures 5-8 and 5-9). Apply the log C method by first deriving the approximate relation for this case that $\alpha_0 - \alpha_2 = 100\mathbf{D}$.

6 Applications of Acid–Base Relations

In this chapter, we treat a variety of acid–base equilibrium situations to give practice in using pH calculations and concepts in a broad span of science.

1. pH Indicators

Titrations constitute a major application of quantitative calculations and these receive a chapter to themselves later. Allied to this is the topic of color-change indicators used to signal end points in titrations, and also used for colorimetric pH determination. These indicators are intensely colored substances whose color changes with gain or loss of protons. The intense color ideally allows use of a quantity so small that only a negligible fraction of acid and base involved in the titration is used to change the color.

A typical indicator is phenol red (phenolsulfonephthalein), a diprotic acid exhibiting two color changes as the phenolic protons are removed or returned. Its structure and α diagram are shown in Figure 6-1. It has $\mathrm{p}K_1^\circ$ 1.5 and $\mathrm{p}K_2^\circ$ 7.9. Just as with other acids, the concentrations of conjugates become equal when $\mathbf{pH} = \mathrm{p}K_a$. In this case, there are two such points at which half the dye is red and half is yellow. The effects of ionic strength can be large, depending upon the charges of the ions formed by the indicator. (See the citric acid example,

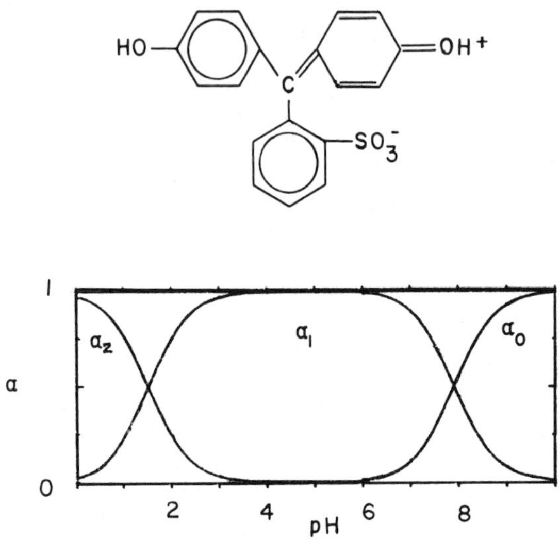

Figure 6-1. Phenol red, structure of acidic zwitterion, H_2A. α diagram: H_2A, α_2, red; HA^-, α_1, yellow; A^{2-}, α_0, red.

Figures 5-10 and 5-11.) This is discussed in more detail for indicators by Laitinen.[1]

The range of color change detectable by the human eye is approximately between the ratios of 10:1 and 1:10 of two colored conjugates. The approximate buffer ratio–**pH** relation of equation (3-7) shows that this will be a range of two **pH** units. Experimental indicator ranges are recorded and vary with the color involved. For example, bromcresol green has a reported range of 3.8–5.4 and phenol red has a range of 6.4–8.0 **pH** units. These are approximate. The lower **pH** change for phenol red is not commonly used and no range is reported for it.

2. Separations

In qualitative and quantitative separations, pH control is often vital to success. We give a few of the many possible examples.

[1]H. A. Laitinen, *Chemical Analysis*, McGraw-Hill, New York, 1960, p. 51 (p. 49 in 2nd ed., 1975).

1. Separations of metal sulfides can be effected by pH control. The product of K_1K_2 for the diprotic acid H_2S [equation set (5-1)] gives the relation between the critical ion S^{2-} and p**H**:

$$[S^{2-}] = K_1K_2[H_2S]/\mathbf{H}^2$$

In saturated solution at room temperature, H_2S is about 0.1 *M*. Thus, putting in the *K* values as well, we obtain

$$[S^{2-}] = 10^{-22}/\mathbf{H}^2$$

for the conditions often used in precipitation. Note that we use only approximate figures because we are comparing large solubility differences under a variety of conditions of temperature and ionic strength. For the more insoluble metal ion sulfides, like Cu^{2+} (group II in the classical H_2S scheme), an acidity of 0.3 *M* **H** is directed. In the above equation, this gives $[S^{2-}]$ about 10^{-21} *M*, which exceeds the solubility products K_{so} for 0.01 *M* ion solutions of these metal ions, which have K_{so} below 10^{-27} while the group III sulfides, like FeS, have K_{so} above 10^{-22} and remain in solution.

2. Amino acid separations and their determinations are of great importance in biochemistry and medicine. Proteins can be hydrolyzed to produce a mixture of amino acids which were polymerized in the protein. Common means of investigating amino acid mixtures are by paper chromatography, ion exchange column chromatography, and electrophoresis. All depend upon the size, charge, and shape of amino acid molecules in the medium used. A clear picture of the effect of pH upon charge for the three types of amino acids is given by the comparative alpha diagrams in Figure 6-2.

If a solution of amino acids is buffered at pH 6 for electrophoresis, glutamic acid will be mainly anionic and migrate toward the positive pole. Lysine, and other basic amino acids, will be cationic and migrate toward the negative pole. Glycine, and the neutral amino acids generally, will be largely zwitterionic and migrate very little. Thus, mixtures can be separated into three groups by this method.

In paper chromatography, the distance traveled, expressed in the R_f fraction, depends on all the properties of the amino acid that affect its solubility in the moving solvent and its affinity for the paper surface. The pH plays some role. For example, lysine shows R_f of 0.8 in a basic solvent, phenol–NH_3, and R_f of 0.18 in butanol–acetic acid media.

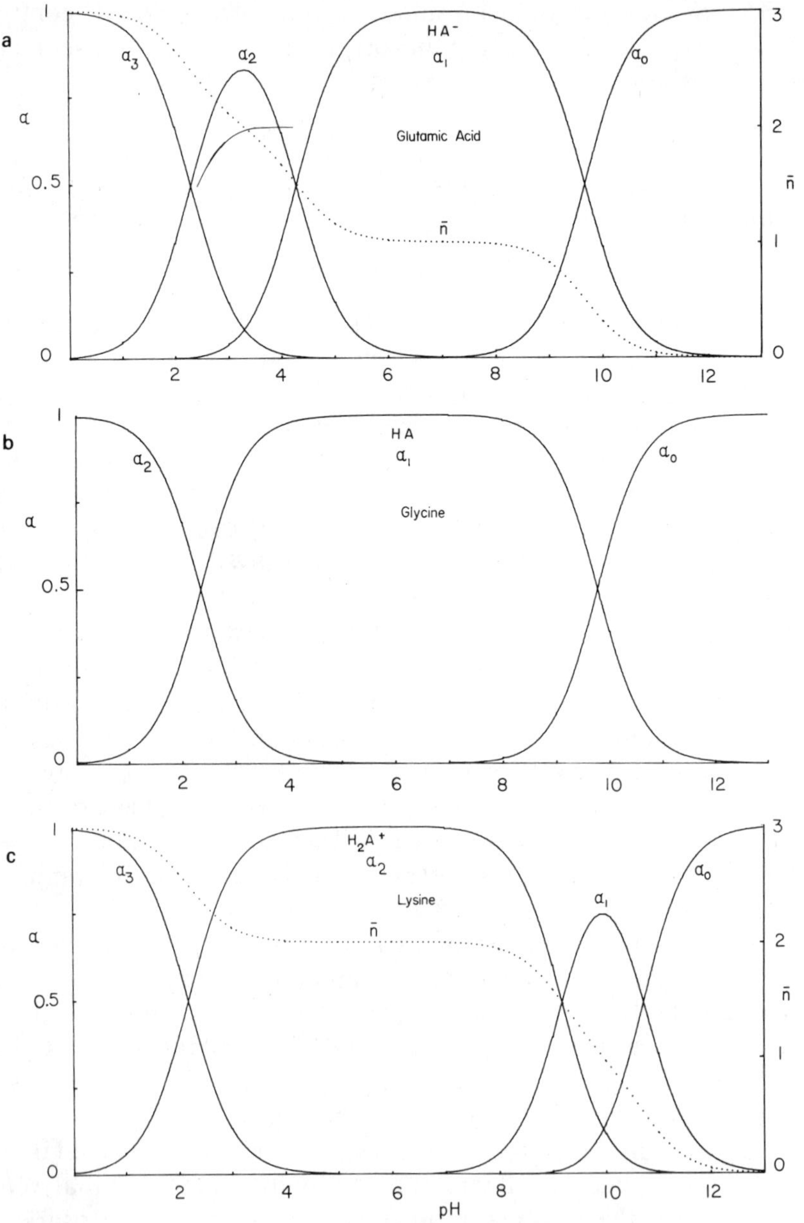

Figure 6-2. Three types of amino acids. (a) Glutamic acid, $I = 0.1\ M$. (b) Glycine, $I = 0$. (c) Lysine, $I = 0.01\ M$. Dotted curves, $\bar{n}$. Solution to Example 3 at $\bar{n}'$ intersection in glutamic acid. Ionic type at **pH** 6 indicated as HA^-, HA, and H_2A^+ for the three cases.

The most extensive amino acid separations can be obtained via ion exchange, a process which depends upon the charge and size of the amino acids and their functional groups. Separation into three groups by charge is easily done. If the solution buffered at pH 6 is passed through an anion exchange resin, usually in the acetate form, the anionic species, like the acid glutamate ion, sorb strongly while the rest run through. These, in turn, can be run through a cation exchange resin in the NH_4^+ form, which holds the cations like H_2Lys^+, allowing the neutral forms to elute.

A technique using very slow passage through long columns of resin with a graded series of buffers to change the pH in small increments has been developed and automated to separate most of the naturally occurring amino acids from each other.

The solvent extraction distribution of amino acids has been studied as a separation process by Craig.[2] In the system he used, 1-butanol–5% aqueous HCl, most amino acids are in cationic forms. Extraction into the butanol layer is favored by larger molecules and by aromatic rings, as in phenylalanine, tyrosine, and tryptophane.

3. Some Physiological pH Effects

Life processes are largely ionic and molecular interactions in aqueous media. Most of the materials involved are pH sensitive and may change their state of protonation as pH shifts. Polymers like the nucleic acids (RNA and DNA) and the proteins (especially enzymes) may change drastically in configuration and ability to function as pH is changed.

The pH of mammalian extracellular fluid is closely controlled by CO_2. This gas can be lost from the lungs, removing acidity, and added by metabolic processes that increase acidity. The balance of this buffer conjugate pair, $H_2CO_3(CO_2)$, HCO_3^-, helps keep the pH in the human arterial blood between 7.35 and 7.45. If through some disorder it falls to 7.0 or rises to 7.8, lethal damage results. All other acid–base pairs in the plasma are forced to assume the ratio determined by the predominant pH of the carbonate system. In general for

[2] L. C. Craig, *Anal. Chem.* **22**, 1346 (1950).

any conjugate pair

$$K_a/\mathbf{H} = [\text{base form}]/[\text{acid form}]$$

Thus, when K_a is smaller than the blood **H**, the protonated form predominates and vice versa. For example, phosphates assume mainly the forms $H_2PO_4^-$ and HPO_4^{2-} as these ratios require at pH 7.4 (see Figure 5-5):

$$K_1/\mathbf{H} = [H_2PO_4^-]/[H_3PO_4] = 10^{-1.95}/10^{-7.4} = 10^{5.4}$$

$$K_2/\mathbf{H} = [HPO_4^{2-}]/[H_2PO_4^-] = 10^{-6.8}/10^{-7.4} = 4$$

$$K_3/\mathbf{H} = [PO_4^{3-}]/[HPO_4^{2-}] = 10^{-11.7}/10^{-7.4} = 10^{-4.3}$$

Phosphates are minor buffering agents in blood but major ones within the cells. The important role of the polyphosphates and their sugar derivatives, especially with adenosine, is described later in this chapter. Their acidities are related to phosphoric acids.

Another important system is NH_4^+–NH_3 with pK_a° 9.24. These are one product of protein catabolism. The acid ion, NH_4^+, can pass through the kidney tubules to be excreted, another means of removing acid from the body. The urine pH varies from about 5 to 7.5 according to the momentary need for elimination of acid in the form of NH_4^+ and also $H_2PO_4^-$. This method, slower than respiration, depends upon recent diet and activity. Citric, lactic, and pyruvic acids and their derivatives are also metabolically important.

Vigorous activity lowers blood pH due to metabolic formation of lactic acid (pK_a° = 3.9) in the muscles. The body reacts to acidity by speeding breathing rate and the removal of CO_2 from the lungs. Conversely, if pH rises, say by digestion of a large meal which draws H^+ into the stomach, relaxation and slower breathing are favored. Detailed discussion of the interpretation and treatment of body pH is given by Woodbury.[3]

The pH controls the precipitation of $CaCO_3$ in the oceans and the formation of the shells of crustaceans. The blood of animals is nearly saturated with respect to $CaHPO_4$ and apatite, $Ca_5OH(PO_4)_3$, the major mineral of bone. A slight increase in pH in certain cells can shift the $H_2PO_4^-$ of blood to more HPO_4^{2-} and PO_4^{3-} to cause deposition of bone.

[3] J. W. Woodbury, in *Physiology and Biophysics*, ed. by T. Ruch and H. Patton, Saunders, Philadelphia, Pennsylvania, 1965, pp. 899–933.

These few examples are a sample of the vast number occurring in the study of life processes. Several of these precipitations are treated in detail in Chapter 11.

4. pH Calculations for Mixtures and Polyprotic Cases

4.1. Mixtures of Two Monoprotic Acid–Base Systems

A consistent mathematical approach to all the problems in this section will greatly simplify these seemingly complex cases. Charge and/or material balancing equations will be the key, leading either to useful equations which are easy to solve exactly, or in some cases to reasonable numerical approximations. The general process is as follows:

1. Obtain an equation of linear terms involving the equilibrium species. Proton balance is usually the best approach, but charge and/or material balancing may be required in complex cases.

2. Use the equilibrium constant expressions to eliminate in step 1 the minor (low concentration) species, retaining **H** and the major species for which we often have good numerical approximations, the analytical concentrations.

3. Solve and check for internal consistency by getting numerical values for all species and checking the material balances.

Example 1. Find **H** in 0.10 M NH_4HCOO solution, ammonium formate. This involves both conjugates of the formic acid and ammonia systems. We perform a proton balance as follows. Write down the hypothetical species added together:

$$0.10\,M\ NH_4^+, \qquad 0.10\,M\ HCOO^-, \qquad H_2O \text{ (assumed constant)}$$

At equilibrium, we have these and add the new species formed, NH_3, HCOOH, H_3O^+, and OH^-. Only these are the result of proton transfers. Therefore, we can set the proton acceptors equal to the donors, a proton balance. The student should convince himself that it is impossible to make an NH_3 without making either HCOOH or H_3O^+, and that it is impossible to make HCOOH without making either NH_3 or OH^-.

Step 1:

$$[\text{proton acceptor products}] = [\text{proton donor products}]$$

$$[HCOOH] + [H_3O^+] = [NH_3] + [OH^-]$$

This perfectly true equation contains four unknowns. We follow the second step to reduce it to a relation of one unknown. Substitute from the conditional molarity equilibrium constant expressions to eliminate $[NH_3]$ and $[HCOOH]$ in favor of $[NH_4^+]$ and $[HCOO^-]$, which we hope will remain approximately 0.10 M. First calculate conditional constants:

$$K_{a_A} = K^\circ_{a_A} f_+/f_- = 10^{-9.24}(0.75)/0.83 = 10^{-9.29}$$
$$= \mathbf{H}[NH_3]/[NH_4^+]$$
$$K_{a_f} = K^\circ_{a_f}/f_+ f_- = 10^{-3.75}/(0.83)(0.775) = 10^{-3.56}$$
$$= \mathbf{H}[HCOO^-]/[HCOOH]$$

(Note that the different charge types have opposing effects here.)

$$[NH_3] = K_{a_A}[NH_4^+]/\mathbf{H}, \qquad [HCOOH] = \mathbf{H}[HCOO^-]/K_{a_f}$$

Step 2. Substitute these into step 1.

$$\mathbf{H}[HCOO^-]/K_{a_f} + \mathbf{H} = [NH_4^+]K_{a_A}/\mathbf{H} + K_w/\mathbf{H}$$

Solve for $\mathbf{H}$,

$$\mathbf{H} = \left[\frac{K_{a_A}[NH_4^+] + K_w}{[HCOO^-]/K_{a_f} + 1}\right]^{1/2}$$

This is a complete, exact equation. To solve, we make approximations. We use the corrected, conditional constants, and we try the approximation 0.10 M for the major ions. Use $K_w = 10^{-13.81}$ previously obtained for 0.10 M ionic strength. We have

$$\mathbf{H} = \left[\frac{10^{-9.29}(0.10) + 10^{-13.81}}{0.10/10^{-3.56} + 1}\right]^{1/2}$$

In both numerator and denominator, the second terms are much smaller than the first terms, so we get

$$\mathbf{H} = (10^{-9.29-3.56})^{1/2} = 10^{-6.42}, \qquad \mathbf{pH} = 6.42$$

Note the value of holding to the exponential form throughout. This result checks in the first step equation. Little of the initial species had to react to form the equilibrium mixture in this case. Be sure to check: This is not always true.

Step 3. $[NH_3] = 10^{-3.87}$, $[HCOOH] = 10^{-3.86}$. We had a rather weak acid, NH_4^+, and a very weak base, $HCOO^-$, resulting in a solution that is slightly acidic. This is the correct **H** at 25°. The student should be able to show that the pH meter should read 6.50 for this solution.

Looking back to the original equation for **H**, we see that the terms we dropped as small have the effect of giving us

$$\mathbf{H} \cong (K_{a_A} K_{a_f})^{1/2}$$

This will often give fair results for solutions of equal concentrations of an acid and a base not conjugate to each other. But it is necessary to check the full equation to be sure.

A clearer picture of the equilibrium situation is obtained by making an α diagram of the systems involved. Figure 6-3 shows us that our major species NH_4^+ and $HCOO^-$ do indeed exist over a wide pH range with α values near one. In a contrasting case, 0.10 *M* ammonium cyanide ($pK_{a_{CN}} = 9.02$, $I = 0.1$ *M*), we see that extensive proton

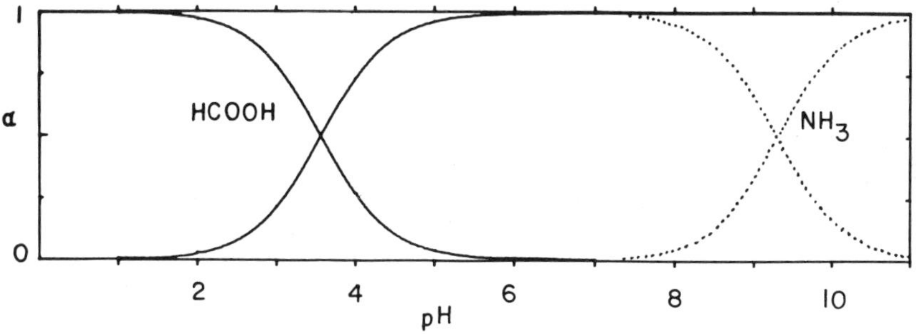

Figure 6-3. Two monoprotic systems for Example 1, formic acid, HCOOH (solid curves), and ammonium (dotted).

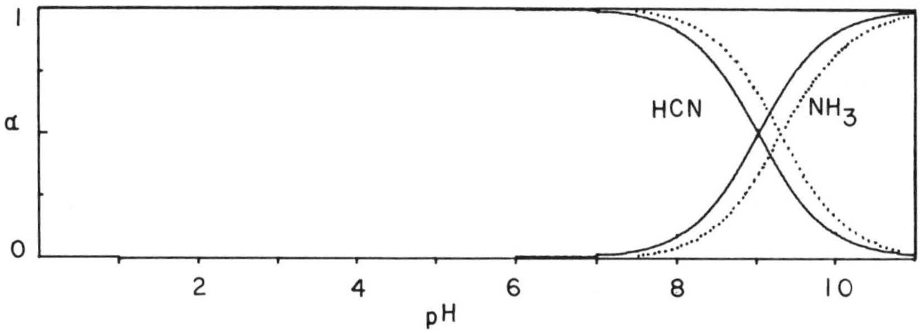

Figure 6-4. Two monoprotic systems for Example 1, HCN (solid curves) and ammonium (dotted).

transfer must occur to reach equilibrium: At no **pH** do both NH_4^+ and CN^- exist together with high α values (Figure 6-4). An algebraic way of testing an equilibrium hypothesis is to check the **pH** required in the K_a expressions for large and small fractions of proton transfer,

$$HX + Y \rightleftharpoons X + HY \tag{6-1}$$

Using the approximate buffer equation (3-5), we get a simple relation:

% completion, (6-1):	0.1%	1.0%	9%	50%
$\mathbf{H} = K_a\mathbf{HX}/\mathbf{X} =$	$1000K_a$	$100K_a$	$10K_a$	K_a
$\mathbf{H} = K_a\mathbf{HY}/\mathbf{Y} =$	$0.001K_a$	$0.01K_a$	$0.1K_a$	K_a

Applying this to our two contrasting cases gives the following sets of **pH** values:

	0.1%	1.0%	9%	50%
$NH_4^+ \rightleftharpoons NH_3$	6.29	7.29	8.29	9.29
$HCOO^- \rightarrow HCOOH$	6.56	5.56	4.56	3.56
$CN^- \rightarrow HCN$	12.02	11.02	10.02	9.02

This shows that equilibrium is reached with little proton transfer near **pH** 6 for the formate, and with much transfer near **pH** 9 for cyanide. Note that equation (6-1) does not require a change in the *ratio* $[NH_4^+]/[CN^-]$, which remains 1:1 regardless of the extent of reaction. The relation $\mathbf{H} = (K_xK_y)^{1/2}$ remains valid unless the **D** terms become large. Check in the complete **H** equation given for ammonium formate in Example 1. For our 0.1 *M* NH_4CN case, the simple equation gives a satisfactory **pH** of 9.15.

4.2. Calculation in Polyprotic Systems

For our first example of a polyprotic case, let us look at the important amino acid glycine. It is a zwitterion molecule,

$$^{+}H_3N{-}CH_2{-}C(=O){-}O^-$$

symbolized HGly. It should be considered a diprotic system involving H_2Gly^+, HGly, and Gly^-. The glycinium ion H_2Gly^+ has the carboxyl protonated, and Gly^- is the glycinate ion. The acidity constants are

$$K_1^\circ = 10^{-2.35} = \frac{(\mathbf{H})(\mathrm{HGly})}{(\mathrm{H_2Gly^+})}, \qquad K_2^\circ = 10^{-9.78} = \frac{(\mathbf{H})(\mathrm{Gly^-})}{(\mathrm{HGly})}$$

(Older literature often contains a K_b which is K_w/K_1, and a K_a which is our K_2.)

Example 2. Find the **H** of 0.10 *M* and of 0.0010 *M* glycine solutions.

Step 1. Initial species are HGly and H_2O. Set equal the new species formed by gain and loss of protons (this is also the charge balance)

$$[\mathrm{H_2Gly^+}] + \mathbf{H} = [\mathrm{Gly^-}] + \mathbf{OH}$$

Step 2. Substitute from the *K* expressions above to get an equation in terms of **H**, [HGly], and the *K*'s:

$$\mathbf{H}[\mathrm{HGly}]/K_1 + \mathbf{H} = K_2[\mathrm{HGly}]/\mathbf{H} + K_w/\mathbf{H}$$

Since ionic strength should be low, we try the K° values and [HGly] of 0.1 *M* in the solved **H** equation,

$$\mathbf{H} = \left[\frac{K_2[\mathrm{HGly}] + K_w}{[\mathrm{HGly}]/K_1 + 1}\right]^{1/2} = \left[\frac{10^{-9.78}(0.1) + 10^{-14}}{0.1/10^{-2.35} + 1}\right]^{1/2} = 10^{-6.08}$$

Step 3:

$$[\mathrm{H_2Gly^+}] = \mathbf{H}(0.1)/10^{-2.35} = 10^{-4.73}$$

$$[\mathrm{Gly^-}] = 10^{-9.78}(0.1)/\mathbf{H} = 10^{-4.70}$$

Thus, [HGly] remains effectively at 0.10 *M* and *I* is below 10^{-4} *M*. Note that slightly more glycinate forms than glycinium and the solution is acidic: Glycine is a better acid than it is a base. The old $K_b = 10^{-11.65}$.

For the 0.0010 *M* case, no terms in the **H** equation are negligible and **H** is $10^{-6.42}$. This gives a $[\mathrm{H_2Gly^+}]$ of $10^{-7.07}$ *M* and a $[\mathrm{Gly^-}]$ of $10^{-6.36}$ *M*. Note the agreement of these results in Figure 6-2.

An important term in amino acid chemistry is "isoelectric point," the pH at which $[\mathrm{H_2Gly^+}] = [\mathrm{Gly^-}]$, which permits the minimum migration in an electric field. The *K* expressions show that this must be exactly at $\mathbf{H} = (K_1K_2)^{1/2} = 10^{-6.06}$. This is approximately, but not exactly, the acidity of the more concentrated glycine solutions.

Amino acids are important enough for us to examine the other two types, acid and basic cases. (See Figure 6-2.) Let us consider 0.010 M solutions of glutamic acid and lysine:

$HOOC—CH(NH_3^+)CH_2CH_2COO^-$, H_2Glu

	H_3Glu^+,		H_2Glu,		$HGlu^-$,	Glu^{2-}
pK:		2.30		4.28	9.67	
I:		0.10 M				

$NH_3^+(CH_2)_4CHNH_2COO^-$, HLys

H_3Lys^{2+},		H_2Lys^+,		HLys,		Lys^-
	2.18		9.18		10.72	
	0.01 M					

Example 3. 0.010 M glutamic acid solution. Let us use the published K_a values given above for 0.1 M ionic strength. (We assume that enough other, inert electrolytes are present to produce this.)

Step 1. Proton balance. Note the Glu^{2-} represents 2 mol of proton transfer:

$$[H_3Glu^+] + \mathbf{H} = \mathbf{OH} + [HGlu^-] + 2[Glu^{2-}]$$

Step 2. Substitute from the K expressions as before to obtain an equation in **H**, $[H_2Glu]$, and K's:

$$\mathbf{H}[H_2Glu]/K_1 + \mathbf{H} = K_w/\mathbf{H} + K_2[H_2Glu]/\mathbf{H} + 2K_2K_3[H_2Glu]/\mathbf{H}^2$$

Let us try to get around the cubic equation by solving for **H** in the quadratic form used above. Multiply through by **H**,

$$\mathbf{H} = \left[\frac{K_w + K_2[H_2Glu] + 2K_2K_3[H_2Glu]/\mathbf{H}}{[H_2Glu]/K_1 + 1}\right]^{1/2}$$

To make a numerical approximation, we can investigate the magnitudes of the terms by making a guess for **H**. Figure 6-2 suggests about **pH** 3–4 for a solution having most H_2Glu. Also, as above with ampholytes, we have the approximate $\mathbf{H} = (K_1K_2)^{1/2} = 10^{-3.29}$. Substitution of this in the last term on the right of our **H** equation shows that term to be negligible compared to the others and we get

$$\mathbf{H} = (10^{-6.28}/3)^{1/2} = 10^{-3.38}$$

Step 3:

$$[H_3Glu^+] = 10^{-3.38}(0.01)/10^{-2.30} = 10^{-3.08}$$

$$[HGlu^-] = 10^{-6.28+3.38} = 10^{-2.90}$$

$$[Glu^{2-}] = 2 \times 10^{-15.95+6.76} = 10^{-8.89}$$

This agrees with the α diagram in Figure 6-2. An appreciable portion of the glutamic acid has formed H_3Glu^+ and $HGlu^-$, so we need a better solution. A second approximation works well in this case. We correct the $[H_2Glu]$ for the other ions it formed and try this value in the **H** equation,

$$[H_2Glu] = 0.010 - [H_3Glu^+] - [HGlu^-] = 0.0079\ M$$

This yields $\mathbf{H} = 10^{-3.40}$. The change is small, about 4% (in the antilogs), so we may take this as a valid solution for **H**.

It may be of interest to compare the graphical method of Chapter 5, equation (5-7),

$$\bar{n}' = C_H/C - \mathbf{H}/C = (0.02/0.01) - \mathbf{H}/0.01 = 2 - 100\mathbf{H}$$

This is the same function for which plotting values are given in Example 1 of Chapter 5. Putting this function upon the equilibrium $\bar{n}$ curve of Figure 6-2 gives the correct solution directly, **pH** 3.40.

The isoelectric **pH** where $[H_3Glu^+] = [HGlu^-]$ (assuming Glu^{2-} to be negligible in this **pH** region) is

$$\mathbf{H} = (K_1K_2)^{1/2} = 10^{-3.29}$$

Example 4. The same procedure applied to lysine, HLys, gives

$$\mathbf{H} = \left[\frac{K_w + K_3[\mathrm{HLys}]}{1 + [\mathrm{HLys}]/K_2 + 2\mathbf{H}[\mathrm{HLys}]/K_1K_2}\right]^{1/2} = 10^{-9.94}$$

The different form results from the different charge types and proton species in the basic, two-amino-group case. The isoelectric **pH** here is 9.95. See the Figure 6-2 correlation with α's.

Example 5. Next we look at the mathematics occurring in an important tetraprotic system, ethylenediaminetetraacetic acid, EDTA. The α diagram is given in Chapter 9, where its uses are investigated further. Find the expected **pH** of a standard 0.020 M Na_2H_2Y solution. For the system at 20°C and ionic strength 0.1 M, published values are

$$H_4Y,\quad H_3Y^-,\quad H_2Y^{2-},\quad HY^{3-},\quad Y^{4-}$$
$$\mathrm{p}K\!:\ 2.07 \qquad 2.75 \qquad 6.24 \qquad 10.34$$

[see equation set (5-1)]. The initial acid–base species put in are H_2O and 0.020 M H_2Y^{2-}.

Step 1. The proton balance among new species produced is

$$2[H_4Y] + [H_3Y^-] + \mathbf{H} = \mathbf{OH} + [HY^{3-}] + 2[Y^{4-}]$$

Step 2. As before, combine the K expressions to obtain substitutions for the minor new species in terms of $\mathbf{H}$ and $[H_2Y^{2-}]$ to give

$$2[H_2Y^{2-}]\mathbf{H}^2/K_1K_2 + [H_2Y^{2-}]\mathbf{H}/K_2 + \mathbf{H}$$
$$= K_w/\mathbf{H} + K_3[H_2Y^{2-}]/\mathbf{H} + 2K_3K_4[H_2Y^{2-}]/\mathbf{H}^2$$

Extract $\mathbf{H}^2$ as in Example 3, leaving some $\mathbf{H}$ terms which may be small on the right. (These terms are the two produced by *two* steps of proton transfer to and from the original H_2Y^{2-} and should be small; see α diagram.) This gives the $\mathbf{H}$ equation

$$\mathbf{H} = \left[\frac{K_w + K_3[H_2Y^{2-}] + 2K_3K_4[H_2Y^{2-}]/\mathbf{H}}{2[H_2Y^{2-}]\mathbf{H}/K_1K_2 + [H_2Y^{2-}]/K_2 + 1}\right]^{1/2}$$

We can use $[H_2Y^{2-}]$ at 0.020 M and a guess as to $\mathbf{H}$ to make a first approximation and to see if any terms are negligible. The α diagram suggests p$\mathbf{H}$ about 4 for the p$\mathbf{H}$ where α_2 predominates. This makes the middle terms of numerator and denominator the largest ones, giving the approximation $\mathbf{H} = (K_2K_3) = 10^{-4.5}$. Using this in the full $\mathbf{H}$ equation gives a second approximation

$$\mathbf{H} = 10^{-4.52}$$

Step 3. Use the K expressions and this $\mathbf{H}$ to calculate the other species concentrations as a check on the EDTA balances:

$$[H_3Y^-] = 3.4 \times 10^{-4}, \qquad [HY^{3-}] = 3.8 \times 10^{-4}$$
$$[H_4Y] = 10^{-9.3}, \qquad [Y^{4-}] = 10^{-5.9}$$

These add up to about 3.5% of the H_2Y^{2-}, so the results are satisfactory to two figures. We assumed 0.1 M ionic strength. The 0.020 M Na_2H_2Y gives only 0.06 of this. Other ions, say 0.04 M NaCl, must be present for these results to be valid.

Example 6. Compare the two contrasting polyprotic systems, 0.010 M triammonium citrate and triammonium phosphate (a hypothetical compound, as we shall see). At ionic strength 0.1 M at 25°, we have the following pK_a values:

for citric acid	2.93, 4.36, 5.74
for H_3PO_4	1.95, 6.80, 11.67

The α curves are found in Figures 5-1 and 5-5. The initial species are 0.03 M NH_4^+, 0.01 M A^{3-}, and H_2O. Proton gain and loss products formed from these give the balance equation,

$$3[H_3A] + 2[H_2A] + [HA] + \mathbf{H} = \mathbf{OH} + [NH_3]$$

Substituting into this from the K expressions [equation set (5-1)] and the K_a for NH_4^+ ($10^{-9.29}$) leads as usual to the complete equation,

$$\mathbf{H} = \left[\frac{K_w + K_a[NH_4^+]}{1 + [A^{3-}]/K_3 + 2\mathbf{H}[A^{3-}]/K_2K_3 + 3\mathbf{H}^2[A^{3-}]K_1K_2K_3}\right]^{1/2} \tag{6-2}$$

The α curves show us that NH_4^+ and citrate ions do exist at α near 1 in the range of **pH** around 8, but that NH_4^+ and PO_4^{3-} cannot be present together in high proportions. For the ammonium citrate, we are led to use the initial concentrations in equation (6-2)

$$\mathbf{H} \cong (3K_3K_a)^{1/2} = 10^{-7.28}$$

This checks in the full equation (6-2) and in the diagram. Figure 6-5 shows the NH_4^+ diagram upon the citric acid curves.

The superimposed α curves of Figure 6-6 show that considerable proton transfer must occur to reach an equilibrium **pH**. Indeed, for

$$NH_4^+ + PO_4^{3-} \rightleftharpoons NH_3 + HPO_4^{2-}$$

we surmise that shifting our solution to the completion of this reaction should provide a more valid situation than the initial concentrations given. That is, we have nearly

$$0.02\ M\ NH_4^+, \qquad 0.01\ M\ HPO_4^{2-}, \qquad \text{and} \qquad 0.01\ M\ NH_3$$

If this is true, the two systems must agree in **H**. For NH_3,

$$\mathbf{H} = K_a[NH_4^+]/[NH_3] = 10^{-9.29}(0.02)/0.01 = 10^{-9.00}$$

We can test the consistency of this with our assumed phosphate distribution, almost all of it as HPO_4^{2-}, 0.01 M. Using K_3, K_2, we have

$$[PO_4^{3-}] = K_3[HPO_4^{2-}]/\mathbf{H} = 10^{-11.67}(0.01)/10^{-9.00} = 10^{-4.67}$$

$$[H_2PO_4^-] = \mathbf{H}[HPO_4^{2-}]/K_2 = 10^{-4.20}$$

Thus, both neighboring species to our major HPO_4^{2-} are under 1% of the $[HPO_4^{2-}]$. Thus, we may safely conclude that our solution is valid. A more satisfactory approach follows.

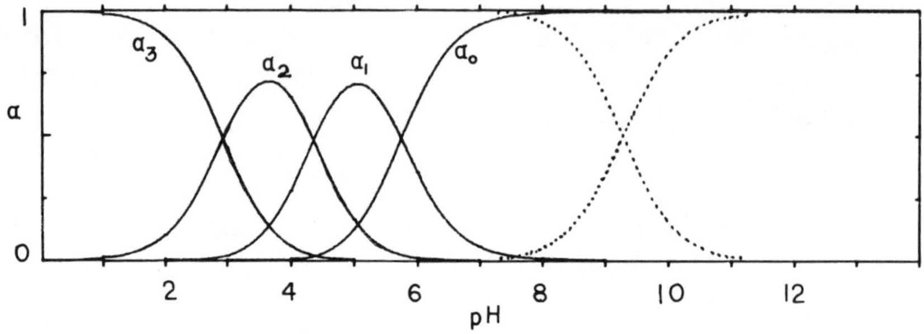

Figure 6-5. The citrate system (solid curves) and the ammonium system (dotted) for Example 6.

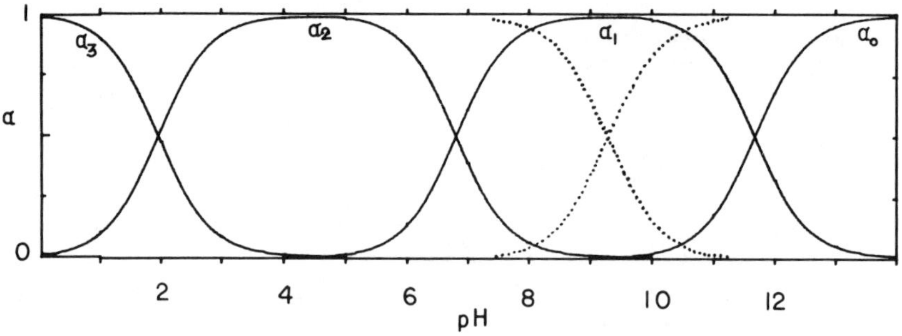

Figure 6-6. The phosphate system (solid curves) and the ammonium system (dotted) for Example 6.

Following the bound proton method used for a mixture in Example 3 of Chapter 5, we can write for the total acid proton concentration

$$C_{\mathrm{H}} = \bar{n}_{\mathrm{A}}(0.01) + \alpha_1(0.03) + \mathbf{D} = 0.030\ M$$

That is, the original solution contains 0.03 M acid protons on ammonium to be redistributed among species. The α diagram shows us that $\bar{n}_{\mathrm{A}}$ remains 1 in the region of interest here, i.e., phosphate has one proton bound to it. This requires α_1 to be 0.67, which we saw above in the K_a expression gives **pH** 9.00. The **D** term is negligible here. This result bears the interesting interpretation that the PO_4^{3-} ion has acted exactly like OH^- here, giving an equivalent amount of NH_3

by proton abstraction from NH_4^+. The ammonia system then determines the **pH**. Checking these results in the full equation (6-2) shows good agreement, giving

$$\mathbf{H} = 1.005 \times 10^{-9}$$

4.3. Effects of pH on Reaction Rates

In general, the rates of solution reactions are proportional to the concentrations of specific species. When these are participants in protonic equilibria, the rates will be found to vary with **H**. One example is seen in the study of ligand exchanges such as

$$Co(NH_3)_5Cl^{2+} + H_2O \rightarrow Co(NH_3)_5H_2O^{3+} + Cl^-$$

The experimental rate is found to depend on the concentrations of the complex ion and on **OH**. One postulated mechanism includes as the rate-determining (slow) step this dissociation of the conjugate base of the complex ion,

$$Co(NH_3)_4NH_2Cl^+ \rightarrow Co(NH_3)_4NH_2^{2+} + Cl^-$$

The concentration of this reactive species is related to the original complex ion through the K_a,

$$K_a = \frac{[Co(NH_3)_4NH_2Cl^+]\mathbf{H}}{[Co(NH_3)_5Cl^{2+}]} = \frac{CB \cdot \mathbf{H}}{[Cpx]}$$

Using the usual α_0 expression for a monoprotic case and $\mathbf{H} = K_w/\mathbf{OH}$, we obtain

$$CB = [Cpx_{tot}]\alpha_0 = [Cpx_{tot}]\left(1 + \frac{\mathbf{H}}{K_a}\right)^{-1} = Cpx_{tot}\left(\frac{K_a \cdot \mathbf{OH}}{K_a \cdot \mathbf{OH} + K_w}\right)$$

When **OH** is low enough to make the K_w the major term,

$$CB = [Cpx_{tot}]\mathbf{OH} \cdot K_a/K_w$$

When K_w is negligible, above a certain **OH**, the rate becomes independent of **OH**. This agrees with the experimental results for a number of reactions of this type.

An examination of this behavior might be helpful in explaining some biochemical reactions. The α diagrams for some biochemically interesting phosphates are shown in Figure 6-7. One can tell at a

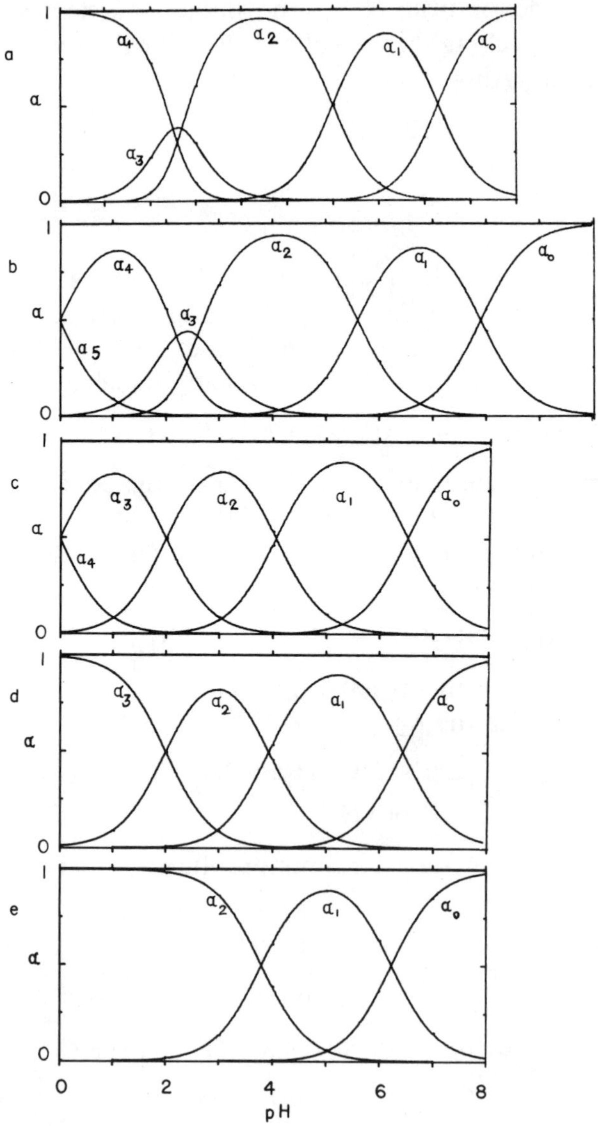

Figure 6-7. Polyphosphate systems. (a) Diphosphoric acid, $H_4P_2O_7$, $I = 0.1\ M$; 25°; pK_a's: 2.5, 2.7, 6.0, 8.3. (b) Triphosphoric acid, $H_5P_3O_{10}$, $I = 0.1\ M$; 25°; pK_a's: 0(?), 2.2, 2.6, 5.6, 7.9. (c) Adenosine triphosphate (ATP), $I = 0.1\ M$; 25°; pK_a's: 0(?), 2(?), 4.06, 6.53. (d) Adenosine diphosphate (ADP), $I = 0.1\ M$; 25°; pK_a's: 2(?), 3.93, 6.44. (e) Adenosine monophosphate (AMP), $I = 0.1\ M$; 25°; pK_a's: 3.80, 6.23.

glance what the major species are at a given **pH**, say serum pH about 7.4. Some **pH** effects can be read directly from these diagrams. For example, a cell, without synthesizing more ATP, could double the concentration of ATP^{4-} ion by changing **pH** from 6.4 to 7.4, changing α_0 from 0.4 to 0.8. More generally, the **pH** dependence of an ion can be obtained from α equations (5-3) and (5-4) as

$$[ATP^{4-}] = \alpha_0[ATP_{tot}] = (\mathbf{H}^4 10^{12.59} + \mathbf{H}^3 10^{12.59} + \mathbf{H}^2 10^{10.59} + \mathbf{H}\cdot 10^{6.53} + 1)^{-1}[ATP_{tot}]$$

From this we may deduce that, if a reaction rate depends on this ion, then at a cell **pH** of 5–6, the first two terms are negligible and the rate depends on the inverse quadratic in **H**. At pH 6–8, the dependence will be on the **H** term, and above **pH** 8, rates should become independent of **H**, and α_0 approaches 1.

Of course reaction rates in biochemical systems may involve more than one acid–base-sensitive material. The rate may go through a maximum at optimum values of α's for the various species.

Further discussion of the acidity effects and equilibria in reactions of these biomolecules may be found elsewhere.[4,5]

5. Summary

While experience may become a guide to rapid decisions as to what may be the major species and controlling equilibria in acid–base mixtures, a reliable indication is always found in a sketch of all the α curves of the systems involved. Then, for relatively simple systems, the algebraic methods of this chapter may be used. For these, and for less obvious cases, the bound **H** method and/or log C diagrams in Chapter 5 may be used. In all cases, the results should be checked for agreement with the α diagrams and the K expressions.

For further readings on this material, see the Selected Reading list on pp. 77–78.

[4] J. N. Lowe and L. L. Ingraham, *An Introduction to Biochemical Mechanisms*, Prentice-Hall, Englewood Cliffs, New Jersey, 1974, Chapter 2.

[5] K. E. Van Holde, *Physical Biochemistry*, Prentice-Hall, Englewood Cliffs, New Jersey, 1971, Chapter 3.

Problems

1. For each solution, give available acid C_H, total polyanion C, and the maximum proton number ratio C_H/C.
   a. 0.10 *M* tartaric acid. This is a diprotic case, H_2A.
   b. 0.10 *M* sodium hydrogen tartrate (NaHA).
   c. 0.10 *M* sodium tartrate (Na_2A).
   d. Mix equal volumes of 0.020 *M* tartaric acid and 0.010 *M* sodium acid tartrate (NaHA).
   e. Mix equal volumes of 0.020 *M* tartaric acid (H_2A) and 0.010 *M* sodium tartrate (Na_2A).
   f. Mix equal volumes of 0.10 *M* citric acid and 0.25 *M* NaOH.

2. Review the categories of polyprotic systems in Chapter 5, and place the solutions in Problem 1 into one category (acid, base, ampholyte, buffer). Estimate the **pH** of each. Tartaric acid has pK_a° values 3.04 and 4.37. Do this by finding or sketching an approximate α diagram and deciding on the **pH** region, what the controlling equilibrium(s) probably are, and what K expressions to use.

3. Prove that these two solutions are indistinguishable:
   a. Mix 10 ml of 0.100 *M* H_3PO_4 + 15 ml of 0.080 *M* NaOH + 5 ml of 0.10 *M* NaCl.
   b. Mix 10 ml of 0.050 *M* HCl + 5.0 ml of 0.14 *M* Na_2HPO_4 + 15 ml of 0.020 *M* NaH_2PO_4.

   Estimate $\bar{n}$ and the **pH** of the solutions from Figure 5-6. Why is this a reliable estimate compared with a case at $\bar{n} = 2.0$?

4. If a 0.10 *M* solution of Na_2HA is acidic, neutral, and basic for three different choices of A^{3-}, what does this tell about $\bar{n}$, the chemical reactions, and the K values? See Example 2 in this chapter.

---

5. What are the ratios of one of the yellow forms to the red form of phenol red at **pH** 2.0 and at **pH** 7.0? Compare results from the approximate ratio equation (3-7) and the α diagram.

6. From the K_a expressions for phenol red, calculate the magnitude of the shift in **pH** with ionic strength change from 0 to 0.1 *M*.

7. If $[S^{2-}]$ is 10^{-24} *M* when PbS starts to form in a given solution and 10^{-8} when MnS starts to form, what is the maximum **H** required to provide these values in saturated H_2S solutions? How could Mn^{2+} and Pb^{2+} be separated?

8. Citric acid and ammonia are often-mentioned metabolism products in the body. What are their actual ionic forms in the blood? Write the K expressions that confirm this.

9. Hydroxylamine, NH_2OH, is a very weak base. The pK_a° at 25° of the conjugate acid, NH_3OH^+, is 5.96. Compare the **pH** of the following three solutions, 0.10 *M*

each: hydroxylammonium chloride, acetate, and cyanide. Sketch rough α curves of the systems and interpret the results chemically.

10. Derive a complete general equation for **H** of solutions containing analytical concentrations A M weak acid and B M weak base, both monoprotic and not conjugate to each other, following methods in Example 1. Put in concentrations of major species assuming a small amount of reaction to form x M conjugate of each starting material.

11. Check the **H** of 0.10 M NH_4CN given in Example 1 in the complete **H** equation, and then solve for the **H** of 0.0001 M NH_4CN.

---

12. Glycine can be titrated either by NaOH or by HCl solutions. Why? Sketch rough titration curves for 10 ml of 0.10 M glycine by 1.00 M HCl and by 1.00 M NaOH. Use one **pH** coordinate.

13. Monosodium glutamate is an important metabolyte and flavoring agent. What **pH** should it produce in 0.010 M solution? See K values in Example 2.

14. Lysine is usually available as the monohydrochloride. Why? What is the **pH** of a 0.10 M solution of this salt?

15. What is the **pH** of a mixture of equal volumes of 0.040 M each monosodium glutamate and lysine monohydrochloride? Look at α diagrams to decide upon the major reactions to use in approximate numerical calculations. Check for contradictions. Show that reaction is under 1%.

16. A citrate buffer has recently been added to the list of standard buffers, Table 2-4. Using published pK_a° values of 3.13, 4.76, and 6.40 and activity coefficients, calculate the pH (activity) expected at 25° and compare with the NBS value.

7 Acid–Base Titration Curves

The familiar curves of pH versus volume of titrant added are displayed and discussed in many texts. Here we shall examine only a few mathematical aspects which follow from our previous treatment and which bear upon the reliability of the end point determination.

1. The Monoprotic Weak Acid–Strong Base Curve

The complete monoprotic **H** equation (3-1) can be adjusted to express the **H** at points during titration if we use the following terms:

C_a = analytical concentration of remaining acid
C_b = analytical concentration of conjugate base, which is also the concentration of base added, in the new volume
$C = C_a + C_b$

Each of these must be calculated at the new volume at each point during titration. Thus, the volume is a concealed variable in these C terms. In equation (3-1), with $C_a = C - C_b$,

$$\mathbf{H} = K_a(C - C_b - \mathbf{D})/(C_b + \mathbf{D})$$

It is convenient to plot a titration curve against the fraction t of titrant added taking the equivalence as 100%. Solve for C_b and divide by C to get the t fraction:

$$t = C_b/C = 1/(1 + \mathbf{H}/K_a) - \mathbf{D}/C = \alpha_0 - \mathbf{D}/C \qquad (7\text{-}1)$$

Now, **H** is the only variable once the acid (K_a) and the initial concentrations and volume are chosen. Curves plotted from equation (7-1) are shown in Figure 7-1. It is convenient to choose values of **H** and calculate t from equation (7-1) rather than doing the reverse. The concealed volume variable sometimes necessitates several successive approximations. For an example, let us find the t at which the p**H** is 3.70 on the titration curve for 50.00 ml of 0.1000 M HA (pK_a 4.00) by 0.1000 M NaOH. The total volume is as yet unknown, but it lies between 50 and 100 ml if we are on the curve. Try 75 ml for a first approximation. We took 5.000 mmol of acid, so C is $5/V$. Equation (7-1) expressed in **H** terms is

$$t = 1/(1 + \mathbf{H}/K_a) - (\mathbf{H} - K_w/\mathbf{H})/C$$

$$C = 5/75 = 0.0667\ M$$

$$t_1 = 1/(1 + 10^{-3.7}/10^{-4.0}) - (10^{-3.7} - 10^{-10.3})/0.0667$$

$$t_1 = 1/3 - 0.003 = 0.330$$

This means that 0.330(50 ml) = 16.5 ml of base should have been added. This predicts a total volume $V = 66.5$ ml and $C = 0.075$. Repeating the calculation with this still gives $t_2 = 0.330$. We have shown that the **D** terms are sometimes small, the α_0 portion being the approximate estimate of titration fraction. The volume and C only affect the **D** term so that, in much of the curve, we get good results for t using rough estimates of V and C.

Figure 7-1 shows the complete range for this equation, since we get strong acid behavior by letting K_a get large so that the first term α_0 becomes 1. We get pure water behavior, or negligibly weak acid, by letting K_a go to zero, which makes the α_0 term zero. Even negative values of t have significance: the fraction of strong acid (equivalent to negative NaOH) that must be added to get those lower p**H** values. Correct points are also obtained for **H** beyond equivalence where the **D** term changes sign and makes t greater than 1.

2. Slope and Buffer Index

The weaker acids in Figure 7-1 have rather small rises at their equivalence points. An expression for this slope can be useful in predicting the feasibility of a given titration and the precision expected.

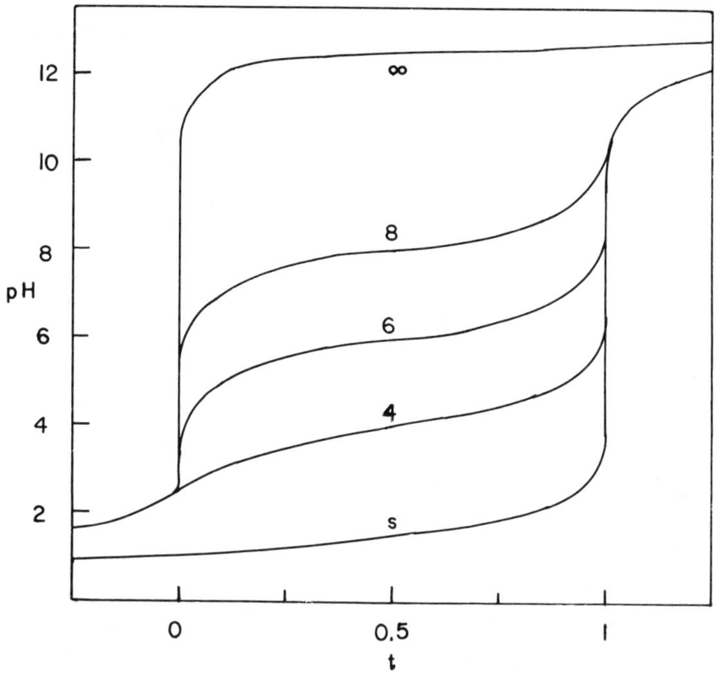

Figure 7-1. Titration curves calculated with equation (7-1). t is the fraction titrated. 0.1 M acids and bases. pK_a values are shown from strong to negligibly weak, pK_a ∞.

This slope also measures the buffering power of the mixture, its resistance to change in **pH**.

We take the equation for C_b and differentiate with respect to **H**,

$$C_b = C/(1 + \mathbf{H}/K_a) - \mathbf{H} + K_w/\mathbf{H}$$

$$\partial C_b/\partial \mathbf{H} = -K_a C/(K_a + \mathbf{H})^2 - K_w/\mathbf{H}^2 - 1$$

C is taken constant for this differentiation. From the definition of **pH** we can derive

$$d(\mathbf{pH}) = -d(\log \mathbf{H}) = -d\mathbf{H}/2.303\mathbf{H}$$

which allows us to convert the previous relation to

$$\partial C_b/\partial(\mathbf{pH}) = \mathbf{B} = 2.303\mathbf{H}[K_a C/(K_a + \mathbf{H})^2 + K_w/\mathbf{H}^2 + 1] \quad (7\text{-}2)$$

B is called the buffer index and expresses the slope as the increment of base required to raise the **pH** by one unit. We transform this to more useful terms using

$$C_b = N_b V_b/(V_0 + V_b)$$

where N_b is the normality of the V_b ml of NaOH added, and V_0 is the starting volume of the weak acid solution. Putting these into equation (7-2), we obtain

$$\mathbf{B} = N_b \frac{\partial(V_b/V_0 + V_b)}{\partial(\mathbf{pH})} = \frac{N_b V_0}{(V_0 + V_b)^2} \frac{\partial(V_b)}{\partial(\mathbf{pH})}$$

This expresses only the left, derivative side of equation (7-2). Now invert and solve for the slope function s in **pH** units per milliliter

$$s = \partial(\mathbf{pH})/\partial V_b = N_b V_0/\mathbf{B}(V_0 + V_b)^2 \qquad (7\text{-}3)$$

Calculation of a slope follows three steps: (1) Use the K_a expression to find the **pH** at the point desired. (2) Use equation (7-2) to find **B** at this point. (3) Use equation (7-3) to find s.

For example, one can find for the pK_a 4 curve above:

At the midpoint, **pH** = pK_a = 4.00, $C = 0.0667$, **B** = 0.0386, and $s = 0.023$ **pH** unit per ml base.

At the equivalence point, **pH** = 8.35 [for 0.05 M NaA by equation (3-4)], $C = 0.05$ M, **B** = 10^{-5}, and $s = 50$ **pH**/ml.

The great difference in slope figure at the midpoint and equivalence point again shows that good buffering obtains when conjugates are present in equal concentrations, and that the slope of the titration curve is steep at the equivalence point unless the acid is very weak. The slopes for a selection of rather strong to rather weak cases are:

pK_a	−1	4	6	8	10
s, **pH**/ml (at end)	1100	50	5	0.5	0.05

As an experimental check, we can look at the actual data plotted in Figure 7-4, a titration of 0.2 M glycine by 0.2 N NaOH which reaches an end point at **pH** about 11.3 and observed slope 0.2 ± 0.05 **pH**/ml base. Using pK_2 9.8 as if this were monoprotic (the pK_1 is far away at 2.4), we obtain $s = 0.2$ **pH**/ml in equation (7-3).

3. Graphical Methods for Location of Titration End Points

Figure 7-1 suggests that there should be little problem in locating the end point of a weak acid in the 0.1 M concentration region if the pK_a is below about 6. Visual indicators and/or **pH** curves serve to establish the steepest slope and inflection point with precision on the order of 1–0.1 % as is commonly desired.

With weaker acids and higher dilutions, even three-figure precision may be difficult to obtain. Either linearizing the pH data or plotting first and second derivative pH curves is used for improving precision. These will be illustrated here for the difficult case of a very weak acid.

3.1. Linearizing Titration Data

In titration of a strong acid, the strong base added removes the acid linearly with respect to the added base. The number of millimoles of strong acid decreases linearly, approaching zero at the equivalence point, and the base increases linearly after the equivalence point. These quantities can be computed from pH readings and plotted to give a V-shaped pair of lines hitting zero at the equivalence point. By definition,

$$\mathbf{H} = 10^{-\mathrm{pH}}/f_+$$

We multiply by the volume of the solution at the point to get the number of millimoles of acid left, which is equal to the original amount, $N_a v_a$, minus the base added, $N_b v_b$:

$$\mathbf{H}V = 10^{-\mathrm{pH}}V/f_+ = N_a v_a - N_b v_b \qquad (7\text{-}4)$$

Here, $V = v_a + v_b$. Now, if the ionic strength, and thus f_+, is constant during the titration, the function $10^{-\mathrm{pH}}V$ should be a linear one versus v_b, the only variable on the right-hand side of equation (7-4).

After the equivalence point,

$$B_t \text{ mmol base} = 10^{\mathrm{pH}-14}V/f_- = N_b(v_b - v_e) \qquad (7\text{-}5)$$

Here, v_e is the volume of base added to reach equivalence, a fixed value, so that the function $10^{\mathrm{pH}-14}V$ increases linearly with v_b after equivalence. Note that the final expressions in both equations (7-4) and

(7-5) are zero at the equivalence point, $v_b = v_e$. Linearization of titration data for strong acid–base cases by this means offers no increase in precision over traditional means of end-point determination in the concentration range about 0.01–1 M. However, this linear plot requires only a few points of pH data for extrapolation. The basic function (7-5) serves as a check if both lines extrapolate to the same end point v_e.

Traditional methods require many readings as the end is approached, and they may be poor as the solutions become more dilute. These and other advantages have brought this method to the attention of analytical chemists. It was introduced by Gunnar Gran in 1952.[1]

3.1.1. Weak Acid Functions

Let us examine the problems of very weak acid titration in some detail. It should be stressed that any method must be tested with known standards to prove that it does what is required—here, produce an end point that agrees satisfactorily with the known equivalence point. The large error that can occur when this is omitted will be illustrated.

What is the linear function of the pH readings in the weak acid case? Taking our complete monoprotic acid equation (3-2) and using

$$C_a = (N_a v_a - N_b v_b)/V \qquad \text{and} \qquad C_b = N_b v_b/V$$

we get

$$\mathbf{H} = K_a(N_a v_a - N_b v_b - \mathbf{D}V)/(N_b v_b + \mathbf{D}V) \tag{7-6}$$

First let us get a satisfactory approximate function.

Case (*a*). In the pH range 4–10, and where C_a and C_b are much greater than **D**. This includes, as discussed in Chapter 3, weak acids of K_a above 10^{-9} and most of the pH curve except very near the start or equivalence in some cases. With the $\mathbf{D}V$ terms negligible in equation (7-6) we can multiply through by v_b,

$$v_b\mathbf{H} = K_a\left(\frac{N_a}{N_b}v_a - v_b\right)$$

[1] Gunnar Gran, *The Analyst*, **77**, 661 (1952).

Again we have v_b as the only variable on the right side. We can transform the left side to pH terms as in equation (7-4),

$$v_b \frac{10^{-\text{pH}}}{f_+} = K_a(v_e - v_b), \qquad v_e = \frac{N_a v_a}{N_b}$$

At constant ionic strength, we can define the weak acid Gran function as

$$W_t = 10^{-\text{pH}} v_b = f_+ K_a(v_e - v_b) \tag{7-7}$$

A plot of W_t vs. the volume of base added v_b should give a straight line hitting zero at equivalence, where $v_e - v_b = 0$. After equivalence, the same base function (7-5) can be used.

In practice, one can multiply through equation (7-7) by any constant for convenient calculation purposes. For example, with pH data between 7 and 10, it may be easier to use antilogs of $(-\text{pH} + 7)$; equation (7-7) can be multiplied through by 10^7. An example plot of the calculated functions for an ideal titration of 0.1 M NH_4^+ ($pK_a^\circ = 9.24$) by 0.1 M NaOH is shown in Figure 7-2. The weak acid titration function W_t and the base function B_t have been multiplied by constants to give one figure before the decimal point.

Case (b). In solutions above pH 9, we may need a complete equation analogous to (7-6) for acids. The **D** term becomes $-$**OH** and we arrange it as

$$\left[\frac{10^{-\text{pH}}}{f_+} N_b v_b - \frac{10^{-14}}{f_- f_+} V - K_a \frac{10^{\text{pH}-14}}{f_-} V\right] = K_a N_b(v_e - v_b) \tag{7-8}$$

The right side shows that the left expression is a linear function of v_b.

Let us examine the degree of approximation we introduce by using the approximate equation (7-7). Figure 7-3a shows calculated plots of Gran functions for two weak acids: acetic acid using 2.0×10^{-5} for K_a and bisulfate ion (HSO_4^-) using 1.0×10^{-2} for K_a. The **D**V terms in the full equation (7-6) have most effect toward the start of titration and these effects increase with acid strength. The full linear form plotted is

$$\frac{10^{-\text{pH}}}{f_+}\left[N_b v_b + V\frac{(10^{-\text{pH}} + K_a)}{f_+}\right] = K_a N_b(v_e - v_b) \tag{7-6a}$$

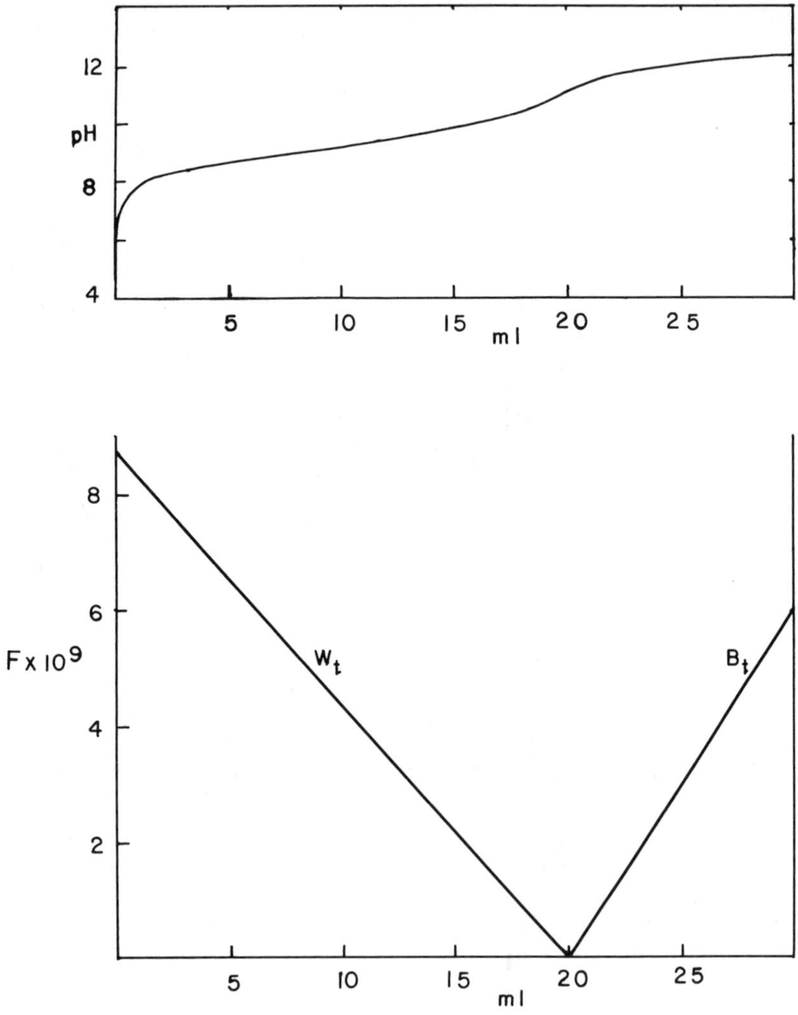

Figure 7-2. Calculated titration curve and its Gran plot for 0.100 *M* NH_4^+ with NaOH, 0.100 *M*. Here **F** is the acid, or the base Gran function.

The left side gives the true linear plot. Clearly, the approximate form (7-7) gives zero at the start when $v_b = 0$. To use equation (7-6a) one must estimate f_+. The uncertainty in pH and K_a will introduce wide deviations toward the start of titration in most cases. Clearly, the approximate Gran plot recommended by most authors requires a rather weak acid for success.

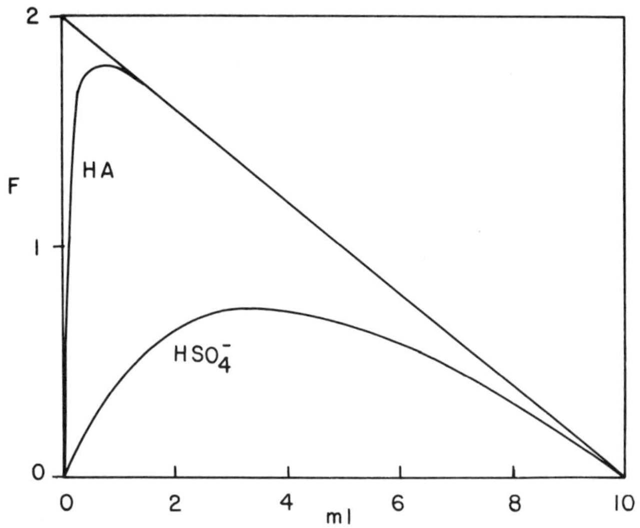

Figure 7-3. Calculated Gran plots by approximate and complete equations for a weak acid (acetic, HA) and a less weak case (HSO_4^-). The complete equation gives the straight line. Normalized to 2 at $v_b = 0$.

3.1.2. Equivalence Point for a Very Weak Acid: Glycine

The functions of equations (7-5) and (7-7) give such fine results for many weak acids that one may be tempted to use them for all cases. A nearly linear plot, but with incorrect end point, results when the conditions in case (a) are not met. Let us examine in detail the titration of 25.00 ml of 0.206 *M* glycine by 0.1902 *M* NaOH. The pH readings are plotted in Figure 7-4a. Ionic strength was held at 0.2 *M* by adding 0.5 g of KNO_3 at the start. The dilution of added base then just balances the added ions. A Gran plot by equations (7-7) and (7-5) is shown in Figure 7-4d and first and second derivative plots are shown in Figures 7-4b and 7-4c, discussed below. These spread Gran-plot end points are not acceptable since the glycine was carefully recrystallized and dried. The expected end point was 27.1 ml, compared with 27.6 (W_t) and 26.5 (B_t) from the linear plots. Let us find the reason for this.

The pH during the last third of the titration is above 10. The $\mathbf{D}V$ terms are not negligible here: At $v_b = 25$ ml, pH 10.612, for equation (7-6) we calculate

$$\mathbf{D}V = \mathbf{OH}(50 \text{ ml}) = 0.02 \qquad \text{and} \qquad N_a v_a - N_b v_b = 0.39$$

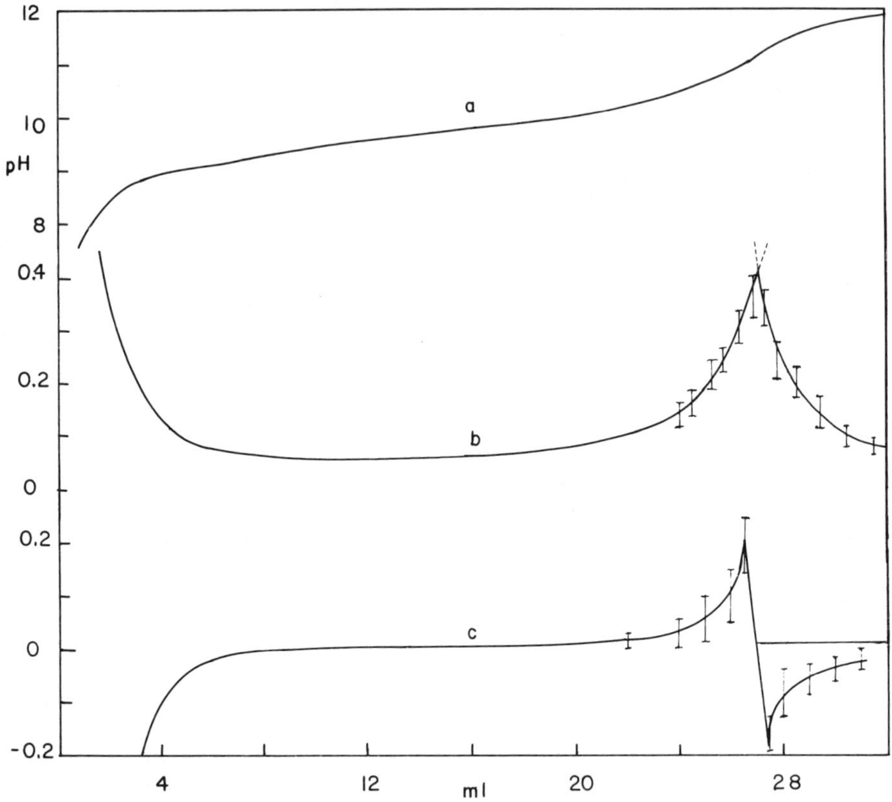

Figure 7-4a–c. The titration of 25.00 ml of 0.206 *M* glycine with 0.1902 *M* NaOH. (a) The pH vs. ml titration curve. (b) First derivative curve, end point 27.1 $\pm$ 0.2 ml. (c) Second derivative curve, end point 27.0 $\pm$ 0.2 ml. The verticals show the estimated uncertainty of the slopes and rates of change of slopes near the end point in (b) and (c).

Thus, we need the full function of equation (7-8) when the equivalence point lies so high in pH. This was calculated with the titration data and plotted as the left line in Figure 7-4e.

For the corrected W_t just plotted and for B_t, we can satisfy ourselves that only the K_2 of glycine need be considered, treating this as a monoprotic titration. Figure 6-2 shows that HGly and Gly^- are the major species in basic solution. pK_1° and pK_2° are 2.35 and 9.78. After equivalence, we must check to see if equation (7-5) is valid when such a strong base as the glycinate ion is present. Its concentration should

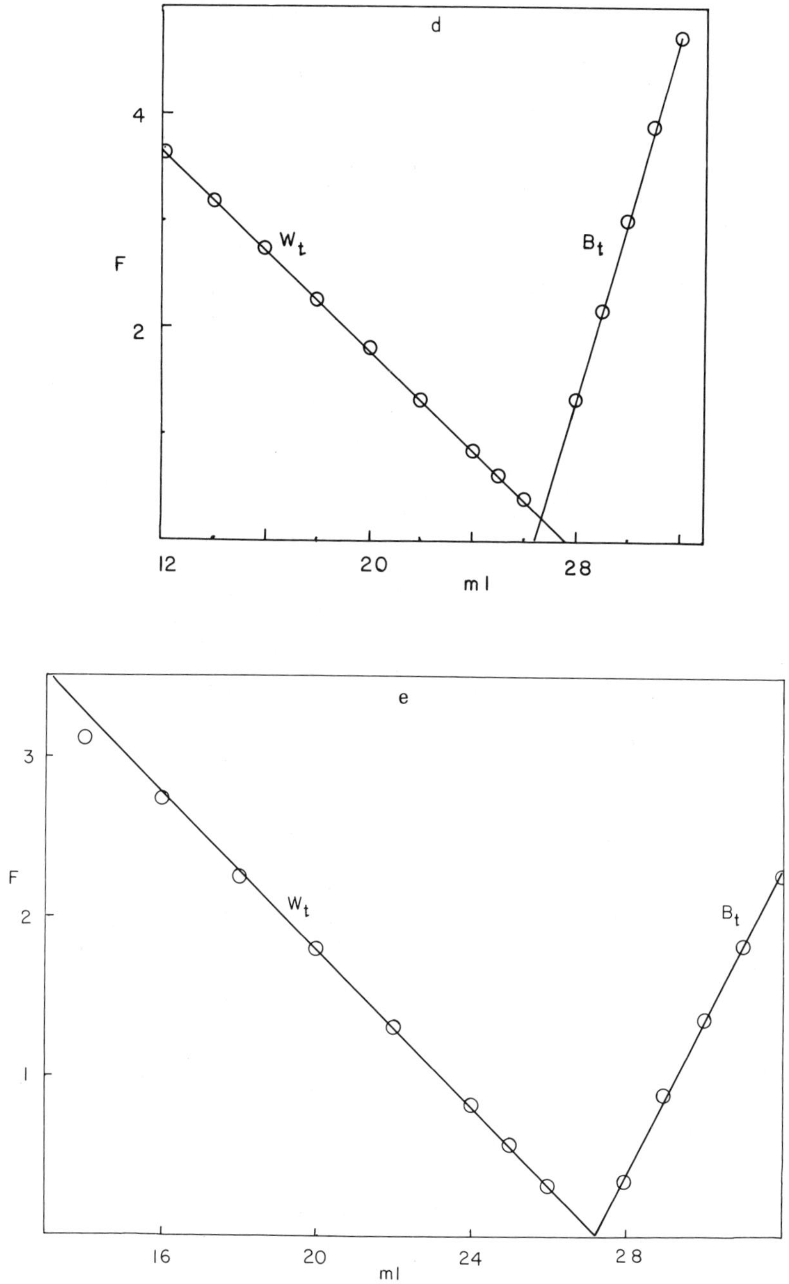

Figure 7-4d, e. (d) Approximate Gran function plots: end points: W_t 27.6 ml, B_t 26.5 ml. (e) Complete equation Gran plots, end point 27.2 ± 0.2 ml.

be about a/V for a mmol of glycine. If a significant amount is removed by

$$\mathrm{Gly^- + H_2O \rightleftharpoons HGly + OH^-}$$

then the total **OH** beyond the equivalence point is the sum of this and the **OH** added:

$$\mathbf{OH} = (v_b - v_e)N_b/V + [\mathrm{HGly}]$$

where [HGly] equals the **OH** from the $\mathrm{Gly^-}$ reaction above. From the K expression this is

$$[\mathrm{HGly}] = f_+\mathbf{H}f_-[\mathrm{Gly^-}]/K_2^\circ$$

Putting this and the usual pH term into equation (7-5), we obtain

$$B_{\mathrm{corr}} = V\left(10^{-14+\mathrm{pH}} - \frac{10^{-\mathrm{pH}}}{K_2^\circ}[\mathrm{Gly^-}]f_-^2\right) = f_-N_b(v_b - v_e) \qquad (7\text{-}9)$$

With $[\mathrm{Gly^-}]$ taken as approximately a/V as explained above, this corrected base function can be computed and plotted vs. v_b. It is quite different from the uncorrected form of equation (7-5) and it is plotted on the right side of Figure 7-4e. This Gran plot now gives reasonable agreement with the weight of glycine and the other plots. This demonstrates the danger in accepting a single method like the plots of equations (7-5) and (7-7) without a known standard and/or an independent analysis.

3.1.3. Derivative Plots

If we can assume that the end point coincides with the equivalence point at the steepest part of the pH curve to the right of Figure 7-4a, then the first derivative (slope) should be a maximum at the end point. Furthermore, the pH curve inflects, so that its second derivative (rate of change of slope) goes through zero at this point. (The equivalence point does not rigorously coincide with this inflection when the pH is not at 7, but the error is smaller than those we are considering in these titrations.)

Numerical values of the slope and its rate of change can be estimated by taking deltas over small volume increments from the titration data or from the smoothed pH curve:

$$\Delta(\mathrm{pH})/\Delta(\mathrm{ml}) = \text{slope}, \qquad \Delta(\text{slope})/\Delta(\mathrm{ml}) = \text{second derivative}$$

When the end point is so poorly defined as in Figure 7-4a, the uncertainty in slopes and second derivative estimates will be large and appreciable scatter is found in their plots, as shown in Figures 7-4b and 7-4c. In this case we find the uncertainty of these two methods to be about the same as the Gran plot value, approxmately 1%, but much better than the vague pH plot itself. For Gran plots, the large number of readings close to the end point is not required. This may give a time saving. If the approximate forms of equations (7-7) and (7-5) are applicable, the Gran method is far simpler than derivative calculations and plotting, which require many good points near equivalence. Automatic potentiometric (pH) titrators are available with provision for reading first and second derivative values. This can speed the process for many routine titrations. But, let us repeat the warning, machine as well as manual methods must be calibrated with known samples and if possible by independent methods of analysis.

3.2. Linearization of pH Data for the Titration of Polyprotic and Mixture Systems

If a single reaction is quantitative within certain pH regions, the equations derived above apply. We treated glycine as a monoprotic acid for the purpose of titration by a strong base. As shown in Figure 5-5, the pH of phosphate systems can be quantitatively described by K_1 alone up to pH 4, by K_2 from about pH 5 to 9, and by K_3 above pH 10. Linearizing equations like (7-5), (7-7), and (7-8) can be derived using v_{e_1}, v_{e_2}, and v_{e_3} for the volumes of titrant added to reach the three equivalence points.

For an illustration, let us examine a titration using standard acid of a solution containing NaOH and Na_2CO_3. Figure 7-5a shows the pH titration curve. The final portion is fairly reproducible (in spite of erratic CO_2 evolution) if one stirs mildly and titrates rapidly.

To linearize the data for the first equivalence point, to determine the amount of hydroxide initially present, we note that the high pH calls for the use of equation (7-9), which deals with a mixture of strong base and a weaker base. The experimentally measured pOH will represent the total of hydroxide remaining from the strong base

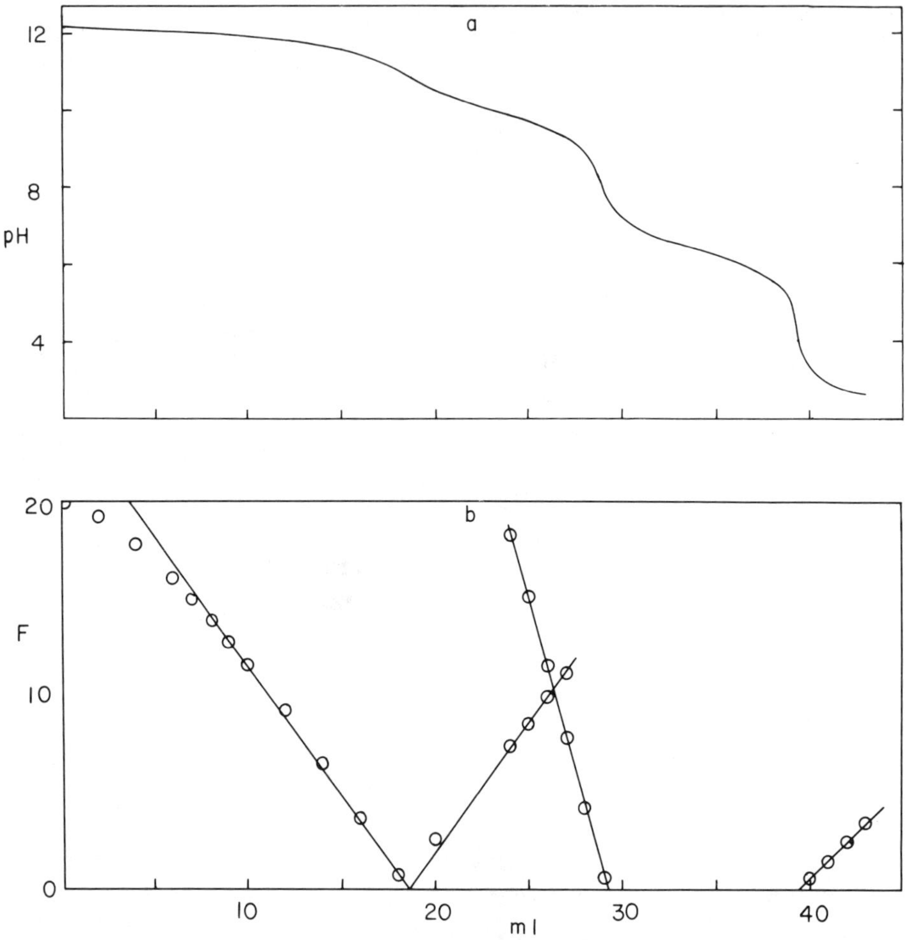

Figure 7-5. Titration of 100.0 ml of hydroxide-carbonate mixture with 0.2062 N_a HCl. (a) pH vs. ml titration curve. (b) Gran function plots revealing three equivalence points: 18.7 ml, and two further segments of 10.3 ml each.

plus any bicarbonate formed by

$$CO_3^{2-} + H_2O \rightleftharpoons HCO_3^- + OH^-$$

which may be appreciable in the pH region here, 11–12. We have

$$(\text{total } OH^-) = (OH^- \text{ left from NaOH}) + (OH^- \text{ equal to } HCO_3^-)$$

$$10^{pH-14}/f_- = N_a(v_{e_1} - v_a)/V + C\alpha_1 \quad (7\text{-}10)$$

where N_a is the concentration of titrant acid, v_{e_1} is the volume of acid added to reach the first equivalence point, v_a is the volume of acid added at any point during titration, V is the total volume of the solution at any point, C is the analytical concentration of total carbonates, and α_1 is the fraction of carbonates present in the form HCO_3^-. At high pH we ignore H_2CO_3 and the K_1 term in the α_1 expression. We have

$$\alpha_1 = \left(1 + \frac{K_2}{\mathbf{H}}\right)^{-1}$$

and

$$C = N_a(v_{e_2} - v_{e_1})/V \qquad \text{or} \qquad C = N_a(v_{e_3} - v_{e_2})/V$$

We put these into equation (7-10) to get

$$\frac{10^{\mathrm{pH}-14}}{f_-} = \frac{N_a(v_{e_1} - v_a)}{V} + \frac{(N_a/V)(v_{e_2} - v_{e_1})}{1 + (K_2/\mathbf{H})}$$

from which we can obtain one possible linear Gran-type function,

$$V\frac{10^{\mathrm{pH}-14}}{f_-} - \frac{N_a(v_{e_2} - v_{e_1})}{1 + K_2 10^{\mathrm{pH}} f_+} = N_a(v_{e_1} - v_a) \qquad (7\text{-}11)$$

To use this, we must first estimate the activity coefficients f_- and f_+ and the equivalence points. Since the latter occur only in the small correction term, they need not be so precisely known as we aim for in the end. Also, we need an estimate of K_2 at the ionic strength used.

Another complication is the appreciable complexing at these concentrations to form $NaCO_3^-$ (ion pairs), thus reducing the carbonate free to participate in the K_2 equilibrium:

$$K_1^\circ = 19 = \frac{(NaCO_3^-)}{(Na^+)(CO_3^{2-})} \qquad \text{(activities)}$$

We calculate that 50% of the carbonate is removed as $NaCO_3^-$, leaving about 1.0 mmol present for the K_2 reaction. With $pK_2 = 9.930$, $f_- = 0.755$, and $f_+ = 0.83$, equation (7-11) becomes

$$V \cdot 10^{\mathrm{pH}-13.878} - [1\ \mathrm{mmol}/(1 + 10^{\mathrm{pH}-10.01})]$$

for the left side experimental Gran function, a linear function of v_a. This has been plotted at the left of Figure 7-5b.

Next we use the K_2 equilibrium in the region between v_{e_1} and v_{e_2} where CO_3^{2-}–HCO_3^- buffering predominates. We have

$$\text{analytical } [CO_3^{2-}] = (v_{e_2} - v_a)N_a/V + \mathbf{D}$$

$$\text{analytical } [HCO_3^-] = (v_a - v_{e_1})N_a/V - \mathbf{D}$$

At pH values below 11, we may try omitting the **D** terms to get

$$\mathbf{H}(v_{e_2} - v_a)/(v_a - v_{e_1}) = \mathbf{H}[CO_3^{2-}]/[HCO_3^-] = K_2$$

One possible linear function is

$$\mathbf{H}(v_{e_2} - v_a) = K_2(v_a - v_{e_1})$$

With $10^{-\text{pH}}$ for the **H** term, we get a function on the left which should be linear when plotted versus v_a. It should rise from zero after v_{e_1}. Another arrangement produces

$$10^{\text{pH}}(v_a - v_{e_1}) = (v_{e_2} - v_a)/K_2 f_+$$

This left side should be a linear function of v_a falling to zero at v_{e_2}. These, and equation (7-4) after v_{e_3}, were plotted in Figure 7-5b to yield the three equivalence points.

Notice that the four lines are more than needed to determine the amounts of hydroxide and carbonate in the starting solution. The final equivalence point and either of the others would totally determine the system. The overdetermination is presented to illustrate the methods and as a desirable means of checking and improving precision. The first equivalence point is least precise, as can be seen in the scatter of points, so its position should be checked by subtracting twice the last segment from the total, 39.3 − 20.6 = 18.7 ml. The plots at the first equivalence point alone only allow a precision of ±0.4 ml. The total base at 39.3 ± 0.1 ml agrees satisfactorily with 39.44 ± 0.04 ml obtained by the highly reliable carbonate method using an indicator and boiling out the CO_2 before final end point adjustment.

Selected Reading

KING, E. J., *Acid–Base Equilibria*, Macmillan, New York, 1965.

KOZAREK, W. J. and FERNANDO, Q., *J. Chem. Ed.*, **49**, 203 (1972).

MACDONALD, T. J., B. J. BARKER, and J. A. CARUSO, *J. Chem. Ed.*, **49**, 200 (1972). This article and the one by Kozarek and Fernando describe applications of Gran's method using the approximate equations only.

RICCI, JOHN E., *Hydrogen Ion Concentration*, Princeton University Press, Princeton, New Jersey, 1952. This book is an important, detailed analysis of the mathematics and interpretation of acid–base phenomena.

ROSSOTTI, F. J. C., and H. ROSSOTTI, *J. Chem. Ed.* **42**, 375 (1965). This article describes Gran's method.

WALTON, H. F., *Principles and Methods of Chemical Analysis*, 2nd ed. Prentice-Hall, Englewood Cliffs, New Jersey, 1964. Chapter 14 gives detailed treatment of titration curve slopes and errors.

Problems

1. Sketch titration curves for 50.00 ml of 0.0100 *M* solutions of formic acid (pK_a 3.7) and for HCN (pK_a 9.2) by 0.0100 *M* NaOH. Calculate the buffer index and the slope at the midpoints and at the equivalence points.

2. Calculate the volume of the 0.01 NaOH in problem 1 during which the indicator would change color in each case if the color change observed occurs over 1.0 pH unit. Express this as a percentage uncertainty in the total titration.

3. 50.00 ml of 0.1 *M* acetic acid, which was also 0.1 *M* in NaCl, was titrated at the pH meter with 0.2005 *N* NaOH. Make a linear plot of the few points taken and find the *N* of the acetic acid giving a ± precision factor which expresses the uncertainty of the method.

v_b, ml	10.00	15.00	20.00	24.00	26.04	27.02	28.0	29.0
pH	4.44	4.79	5.22	6.02	11.38	11.60	11.72	11.87

8 | Metal Ion–Ligand Systems

Stepwise equilibria between Lewis acids, here metal ions, and Lewis bases, here ligands, are mathematically quite like the preceding protonic equilibria. Water is the most concentrated ligand in most solutions. The competition of other ligands for the metal ions is important in the understanding of solution species distributions which play a role in many areas of chemistry, biochemistry, and geochemistry.

The solvent water equilibrium with **H** and **OH** adds terms to the material balance in the protonic equilibrium expressions. Other than that, the formulation is the same for these metal ion cases. We shall show how both are formulated in the formation direction.

1. General View of M–L Systems

Recently experimental data have made it clear that association or complexing between ions of opposite charge is so common in solution that its *absence* is remarkable. Earlier, chemists held the view that most saltlike compounds are in completely ionic form in dilute water solutions. Now, this seems true only for singly charged alkali metal halides, nitrates, and perchlorates and a few other cases. We need not specify the mechanism of the complexing, whether it be covalent, ion pairing, etc., but we take the term to mean any association that can be quantitatively treated by our equilibrium constants. Clearly this is partly a matter of degree, since all oppositely charged

Table 8-1. Formation Constants K_1° for the First Step of Complexing

Ions	K^+, SO_4^2	Ca^{2+}, SO_4^2	Fe^{3+}, Cl^-	Ce^{3+}, Cl^-	Hg^{2+}, Cl^-
K_1°	7	190	30	3	5.5×10^6 [a]

[a] At $I = 0.5\,M$.

ions attract each other to affect their mobilities and activities. Some examples are given in Table 8-1.

Let us illustrate the significance of a $K_1^\circ = 7$ by calculating the fraction complexed in 0.1 M potassium sulfate solution. From equation (2-4) of Chapter 2, we get $I = 0.3$, which leads to $f_{2-} = 0.25$ from the Davies equation or chart. We assume that the other two f factors cancel. We have

$$K_1^\circ = 7 = \frac{[KSO_4^-]f_-}{[K^+][SO_4^{2-}]f_+ f_{2-}} = \frac{x}{(0.2) - x)(0.1 - x)(0.25)}$$

where x is the concentration of the complex formed. Solving for x yields

$$[KSO_4^-] = 0.024, \qquad [K^+] = 0.176, \qquad [SO_4^{2-}] = 0.076$$

Thus, 12% of the K^+ and 24% of the SO_4^{2-} originally added are in associated form. This significantly affects the solubility of $CaSO_4$ in the case studied in Chapter 10.

Recent quantitative investigation of complexing in systems of weak interactions has led to an understanding of previously mystifying situations in sea water and biological fluids. The range of constants in Table 8-1 is large. The chloride concentration in $HgCl_2$ solutions is so low that AgCl may not precipitate when dilute $AgNO_3$ is added. Furthermore, no simple relation to charge type is apparent. Some theoretical sense in explaining contrary effects in Lewis and protonic acid–base interactions has come from "hard–soft" acid–base ideas.[1]

Table 8-2 lists formation constants, K_1, or overall β where only those are available, as log values. The softness of acids and bases in this approach seems to correlate with polarizability: large size, low

[1]R. G. PEARSON, ed., *Hard and Soft Acids and Bases*, Dowden, Hutchinson, and Ross, Stroudsburg, Pennsylvania (1973).

Table 8-2. Values of the Logarithm of the Formation Constant, $\log K_1$ at 25°, Estimated for $I = 0$[a]

Ion	OH^-	F^-	Cl^-	NH_3	Br^-	I^-	SCN^-	CN^-	$S_2O_3^{2-}$
H^+	14	3.2	—	9.2	—	—	—	9.2	2
Mg^{2+}	2.6	1.8	—	0.2	—	—	—	—	1.8
Ca^{2+}	1.4	1	—	−0.2	—	—	—	—	2.0
Al^{3+}	8.9	6.2	—	—	—	—	0.4	—	—
Cr^{3+}	10.2	5.2	−1	—	−2.6	—	3	—	—
Fe^{3+}	11.8	5	1	—	0.6	2	3	$(31, \beta_6)$	2
Mn^{2+}	3.4	—	0.1	0.8	—	—	—	—	—
Fe^{2+}	4.5	1	< -0.3	1	—	—	—	$(24, \beta_6)$	2
Co^{2+}	5.1	—	−0.3	2	−0.1	—	1.7	$(19, \beta_6)$	2
Ni^{2+}	3.4	1	−0.5	3	−0.3	—	1.8	$(30, \beta_4)$	2
Cu^{2+}	7.2	1	1	4	—	—	2.3	—	—
Sn^{2+}	10.1	5	1	—	1	—	1	—	—
Pb^{2+}	6.2	1	1.6	1	2	2	—	—	2
Bi^{3+}	12.4	—	2	—	—	3.6	1	—	—
Zn^{2+}	5.2	1	0.2	2.4	−0.5	−2	1	$(17, \beta_4)$	2
Cd^{2+}	5.0	0.5	2.0	2.6	2.2	2.7	1	5	4
Hg^{2+}	10.3	1.6	6.7	8.8	9	13	9	18	30
Ag^+	2	0.4	3	3	4	8	5	$(20, \beta_2)$	9

[a] The order is from hard to soft in going from left to right (OH^- to $S_2O_3^{2-}$) and from top to bottom (H^+ to Ag^+).

charge, and deformable electron clouds. Hardness correlates with small size, high charge, and tight electron clouds, often *s* or *p*, at the surface of the ion. There is no simple correlation with protonic acidity in water. This hard–soft property seems superimposed upon other factors. The Lewis bases are arranged empirically in hard to soft order for the donor atom:

$$\text{(hard)} \quad \text{F, O, N, Cl, Br, I, C, S} \quad \text{(soft)}$$

Here, the O, for example, may be in OH^-, $RCOO^-$, ROH, etc. The order is only roughly constant. The acid ions are arranged into three groups: hard, borderline, and soft as ordered in Table 8-2.

In Table 8-2, we can see that the hard acids form more stable, higher K, complexes with hard bases and similarly for the soft acids with soft bases. The trends across the table are reversed roughly as we move from top to bottom. Examples are: hard–hard, AlF^{2+} (log K_1 6.2); soft–soft, $HgCN^+$ (log K_1 18); hard–soft, HgF^+ (log K_1 1.6). Often, but not in all cases, hard–soft and soft–hard combinations are weaker interactions.

Care must be taken to differentiate between stability and ease of reaction. The term *stability* refers to equilibrium position, extent of possible reaction, and high vs. low K values. It has no simple relation to speed or ease of reaction. The terms *labile* and *inert* refer to rates. Most of the complexes in Table 8-2 form as rapidly as the reagent solutions are mixed. However, those of Cr^{3+} form, and exchange, very slowly. These are called inert complexes. Most of our work in this text concerns labile, rapidly reacting complexes. Before assuming that a transition metal complex is labile, one should consult the literature of inorganic chemistry to be sure. Titrations (in the cases of Cr^{3+} and Co^{3+} by EDTA) require special provisions to overcome slow reactions, for one important example. To know whether given ions may form complexes and to what extent, one can consult the volumes of *Stability Constants*.[2] Naturally they list only those which have been quantitatively investigated. However, analogy can yield insight into likely behavior: For example, we can deduce from Table 8-2 that Cd^{2+} and Hg^{2+} should bind well to sulfide or amine positions in organic molecules.

2. Useful Relations Derived from Sets of Step Equilibria

We have shown that all mononuclear equilibria have similar equilibrium constant expressions. Now, we can show that general equations can be found to give the fractions of the species present as a function of uncomplexed ligand or proton concentration alone. Also expressable in terms of the one variable is the $\bar{n}$ function, the ligand number, proton number, or degree of formation. The $\bar{n}$ is the average number of coordinated ligands or protons per central group

[2] L. G. Sillen and A. E. Martell, *Stability Constants of Metal-Ion Complexes*, Special Publication No. 17, 1964, and No. 25 (Supplement), 1971, The Chemical Society, London.

in the system at equilibrium. These equations lead to clear-cut, complete-equation graphical solutions to equilibrium problems which facilitate understanding of complex systems and methods of computer programming for more rapid solutions.

(*i*) Fractional distribution of species, alpha equations. The fraction of a species is its concentration ratio to the total. For M–L systems, C_M will be the total (analytical) concentration of the central group (usually metal ion). In the case of set (5-1), this is

$$C_M = [M] + [ML] + [ML_2] + [ML_3] + [ML_4]$$

This is the sum of all forms of M assumed to be present. This is called the *material balance* on M (or, indirectly, the mass balance). The fraction in the form of the "free" aquo ion is then

$$\alpha_0 = [M]/C_M = (1 + [ML]/[M] + [ML_2]/[M] + [ML_3]/[M] + [ML_4]/[M])^{-1} \tag{8-1}$$

From the equilibrium expressions of set (5-1) we can substitute for each ratio in (8-1) an expression in terms of [L] just as in the α derivations in Chapter 5:

$$\alpha_0 = (1 + K_1[L] + K_1K_2[L]^2 + K_1K_2K_3[L]^3 + K_1K_2K_3K_4[L]^4)^{-1} \tag{8-2}$$

And, as before, the rest of the α's can be expressed in terms of this α_0:

$$\begin{aligned}
\alpha_1 &= [ML]/C_M = K_1[M][L]/C_M = \alpha_0 K_1[L] \\
\alpha_2 &= [ML_2]/C_M = K_1K_2[M][L]^2/C_M = \alpha_0 K_1K_2[L]^2 \\
\alpha_3 &= [ML_3]/C_M = \alpha_0 K_1K_2K_3[L]^3 \\
\alpha_4 &= [ML_4]/C_M = \alpha_0 K_1K_2K_3K_4[L]^4 \\
&\vdots \\
\alpha_N &= [ML_N]/C_M = \alpha_0 \beta_N[L]^N
\end{aligned} \tag{8-3}$$

where N is the maximum value of n, the number of ligands on M. The important meaning of these equations and those in set (5-1) is that in systems where the equilibrium constants are known, determination of the uncomplexed ligand (or proton) concentration at equilibrium will be sufficient information to tell the distribution of M (or A) among all the possible forms it can take. This is the basis of many equilibrium diagrams we shall use. It may seem remarkable that C_M is not needed for this. In reality, C_M (or C_A in the protonic cases) does affect [L] (or **H**), so that this master variable alone describes the situation at equilibrium.

(*ii*) $\bar{n}$, the ligand number, or degree of formation. This is analogous to the proton number in Chapter 5. Again, this is a useful function because it represents a quantity that can sometimes be found experimentally. It is the average number of bound ligands per metal ion. The bound ligand concentration must be the total put in, C_L, minus that left unbound at equilibrium, [L]. This leads to the material balance and equilibrium expressions for $\bar{n}$ just as in Chapter 5:

$$\bar{n} = \frac{C_L - [L]}{C_M} = \frac{[ML] + 2[ML_2] + 3[ML_3] + 4[ML_4]}{C_M} \tag{8-4}$$

From the α expressions of set (8-3), this can be written

$$\bar{n} = \frac{C_L - [L]}{C_M} = \alpha_1 + 2\alpha_2 + 3\alpha_3 + 4\alpha_4 \tag{8-5}$$

For certain purposes, it is useful to put in the α expressions and get general expressions in terms of [L]. We use the symbols for overall constants (e.g., $\beta_3 = K_1K_2K_3$). We have

$$\bar{n} = \frac{\beta_1[L] + 2\beta_2[L]^2 + 3\beta_3[L]^3 + \cdots + N\beta_N[L]^N}{1 + \beta_1[L] + \beta_2[L]^2 + \beta_3[L]^3 + \cdots + \beta_N[L]^N}$$

Multiply both sides by the right denominator and collect terms in powers of [L] to get

$$\bar{n} = (1 - \bar{n})\beta_1[L] + (2 - \bar{n})\beta_2[L]^2 + \cdots + (N - \bar{n})\beta_N[L]^N$$
$$\bar{n} = \sum_{n=1}^{n=N} (n - \bar{n})\beta_n[L]^n \tag{8-6}$$

Measurement of n experimental values of $\bar{n}$ at different [L] will give n equations in n unknowns, the K values. Thus, equation (8-6) provides a starting point for determination of equilibrium constants from data.

Equation (8-5), like (5-7) for protonic cases, has two expressions for $\bar{n}$, one from material balance, and one from equilibrium expressions (the α's). Intersection graphical solutions to problems can be achieved in the same fashion as we did in Chapter 5, using $\bar{n}$ plots.

3. General Features of Species Distributions

Here we examine the regularly recurring features of the α fractions of species in step equilibrium when plotted against log [L] (or log **H** for protonic cases). Even very roughly sketched α diagrams can be helpful in understanding equilibrium situations and in making valid approximations to achieve numerical solutions to problems. Therefore, it is valuable to develop facility in recognizing the types of curves that always result from various sets of equilibrium constants. These are identical for protonic and M–L cases. In Chapter 5, we plotted p**H** from left to right so that **H** is high at the left, and **OH** at the right of those diagrams. Here we shall use [L] increasing left to right.

(*i*) $N = 1$. As discussed in Chapter 4, the α_0 and α_1 curves for these systems are invariant in shape. The difference here is only the use of formation, instead of dissociation direction. From equation (8-2), we get equations for α analogous to (4-3),

$$\alpha_0 = (1 + K_1[\mathrm{L}])^{-1}, \qquad \alpha_1 = (1 + 1/K_1[\mathrm{L}])^{-1}$$

As always, $\alpha_0 + \alpha_1 = 1$. For example, take $K_1 = 10^4$, and calculate α values at various [L], as before, at $[\mathrm{L}] = 0.1K_1$, K_1, $10K_1$, etc. This produces Figure 8-1, with the two α curves crossing at $\log[\mathrm{L}] = -\log K_1$, where $\alpha_0 = \alpha_1$. In single-K systems, $\bar{n} = \alpha_1$.

(*ii*) $N = 2$. More variety occurs here. We can observe the influence of various spacings of the K values by plotting the curves for some examples:

	$\log K_1$	$\log K_2$	ratio K_2/K_1
(a)	6	2	10^{-4}
(b)	4.5	3.5	0.1
(c)	4.15	3.85	0.5
(d)	4	4	1
(e)	3.5	4.5	10

From equation (8-2) we can write the α's for case (a) as

$$\alpha_0 = (1 + 10^6[\mathrm{L}] + 10^8[\mathrm{L}]^2)^{-1}$$

$$\alpha_1 = 10^6[\mathrm{L}]\alpha_0$$

$$\alpha_2 = 10^8[\mathrm{L}]^2\alpha_0$$

We can choose suitable values of [L], say $10^{-7}, 10^{-6}, \cdots, 10^{-1}$, and

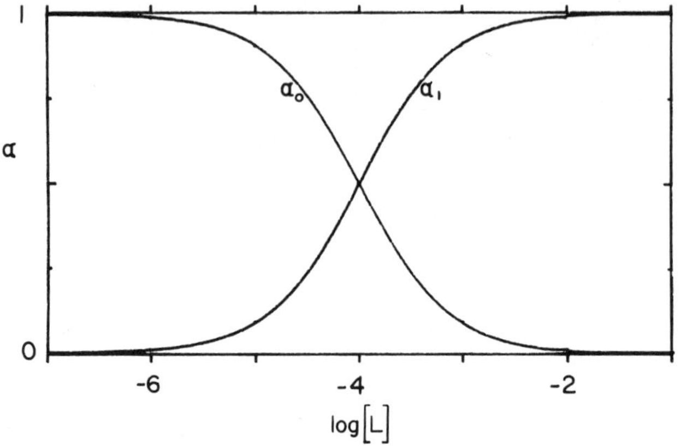

Figure 8-1. The α curves for a single-K system, ML. Log $K_1 = 4$.

calculate enough α's to get the curves as shown in Figure 8-2. Cases (b)–(e) are plotted in the same way.

Case (a) is simply two adjacent sections each like Figure 8-1. The values of K_1 and K_2 are so far apart that all three species are never present in appreciable proportions at any value of [L]. This will be characteristic of all cases having K values higher than about 4 log [L] (or p**H**) units apart.

When K values are closer together as in (b) and (c), α_1 never rises near the full (100%) value. In (d), all three α curves must intersect at $\alpha = 0.33$ at $[L] = K^{-1}$. (For proof see K expressions.)

Finally, in case (e), when K_2 is greater than K_1, α_1 is never the major species; [M] and/or $[ML_2]$ are greater than [ML] at all values of [L]. This is not common. The systems Ag^+–NH_3 and Hg^{2+}–Cl^- seem to be examples. The set of $\bar{n}$ curves, $\alpha_1 + 2\alpha_2$, is shown for comparison in Figure 8-2.

By referring to the K expressions, we can see where intersections must occur, where species become equal to each other. When [M] = [ML], $\alpha_0 = \alpha_1$ and $[L] = 1/K_1$. Similarly, $\alpha_1 = \alpha_2$ at $[L] = 1/K_2$. Then, $\alpha_0 = \alpha_2$ at $[L] = 1/(K_1K_2)^{1/2}$. The maximum of α_1 can be found by differentiation of the expressions for α_1^{-1}, as in Chapter 5, to get $[L] = 1/(K_1K_2)^{1/2}$. In the α_1 expression this gives

$$\alpha_1 = [1 + 2(K_2/K_1)^{1/2}]^{-1}$$

Since $\alpha_0 = \alpha_2$ here, and $\alpha_0 + \alpha_1 + \alpha_2 = 1$, we get

$$\alpha_0 = \alpha_2 = (1 - \alpha_1)/2 = [(K_1/K_2)^{1/2} + 2]^{-1}$$

(iii) $N = 3$ *and larger values.* No really new types of curves are found, merely combinations of what are seen in Figure 8-2. The $(N - 1)$ α curves between the end ones, α_0 and α_N, go through maxima at [L] values and α heights which depend on the spacings of K values. Some cases for three and six constants, four and seven species, are shown in Figures 8-3 and 8-4.

Figures 8-3a and 8-3b show the effect of regularly decreasing K values: (a) log K of 8, 5, and 2; (b) log K of 5, 4.5, and 4. The first acts almost like three separate single systems (like H_3PO_4, Figure 5-5). The $\bar{n}$ curves reflect the interaction in Figure 8-3b by showing no flat segments of nearly constant $\bar{n}$ as does Figure 8-3a. Figure 8-3c shows one widely spaced pair and the second pair in inverse order, log K of 7, 3, and 4. A completely inverse set in Figure 8-3d shows how the first and last species predominate. The log K values are 4, 5, and 6.

For six-complexing, Figure 8-4a shows an evenly spaced set, log K of 8, 7, 6, 5, 4, and 3. Figure 8-4b has log K of 7, 6, 5, 5, 4, and 3. Figure 8-4c is in reverse order, with values of log K of 3, 4, 5, 6, 7, and 8. Only α_0 and α_6 become visible in this plot. A log α plot is needed to show the small intermediate α_1, α_2, α_3, α_4, and α_5.

For an actual case, let us examine the system Al(III)–F^-. The rather even spacing of the log K values produces the result shown in Figure 8-5a, similar to Figure 8-4a. The constants published for 25° at 0.5 M ionic strength are, log K_1 to log K_6,

6.164 5.053, 3.91, 2.71, 1.46, 0.

Note that a value zero means K_6 is one, and that only one figure was obtainable experimentally.[3]

Equation (8-2) for computing α_0 is now

$$\begin{aligned}\alpha_0^{-1} &= 1 + 1.46 \times 10^6[F^-] + 1.65 \times 10^{11}[F^-]^2 \\ &\quad + 1.35 \times 10^{15}[F^-]^3 + 6.9 \times 10^{17}[F^-]^4 \\ &\quad + 2.0 \times 10^{19}[F^-]^5 + 2.0 \times 10^{19}[F^-]^6\end{aligned} \tag{8-7}$$

[3]E. L. King and P. K. Gallagher, *J. Phys. Chem.* **63**, 1073 (1959).

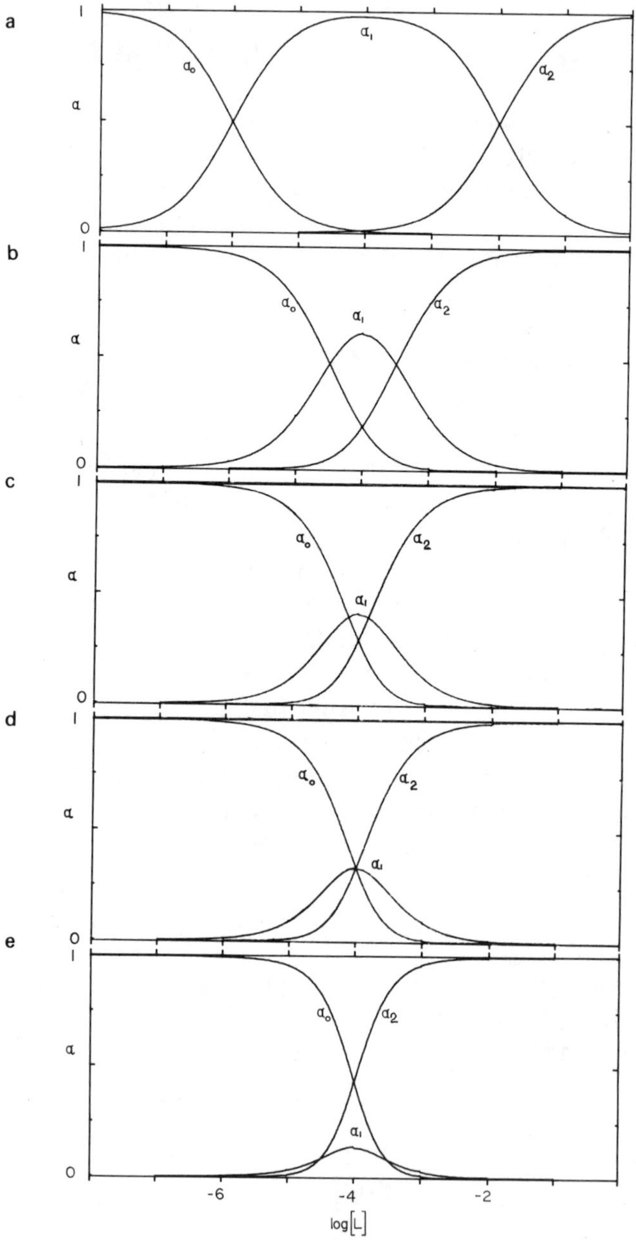

Figure 8-2. The α and $\bar{n}$ curves for systems of two constants, M, ML, ML_2 present. See text for constant values.

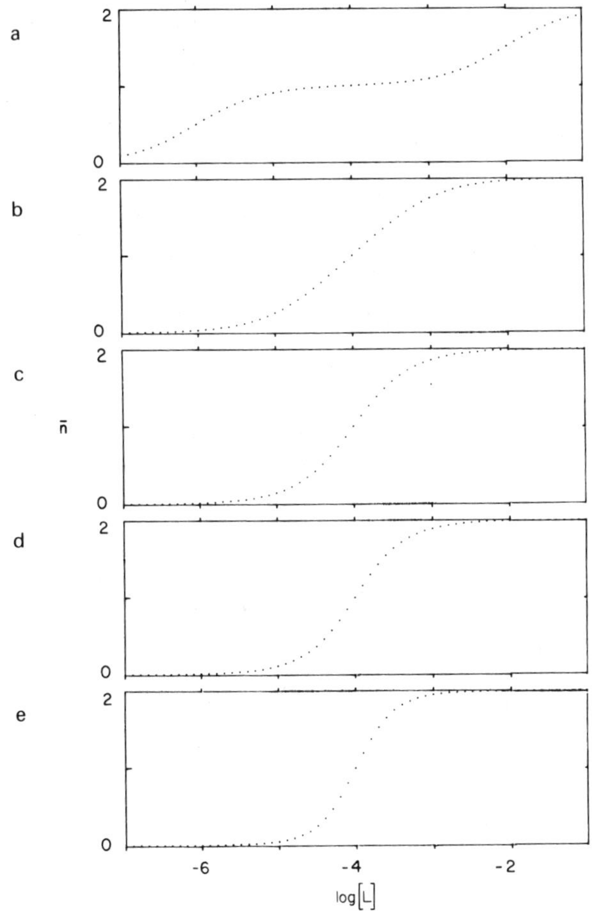

Figure 8-2 continued.

A study of the types of α curves in Figures 8-2–8-5 reveals general features when the K_n values decrease:

1. As the separation of adjacent K_n values increases, so does the height of the maximum between them.

2. Adjacent α_n curves cross at $[F^-] = 1/K_n$, where n is the high n value. For example, α_0 crosses α_1 at $\log [F^-] = -6.16$ for the same reasons given above under the $N = 2$ cases.

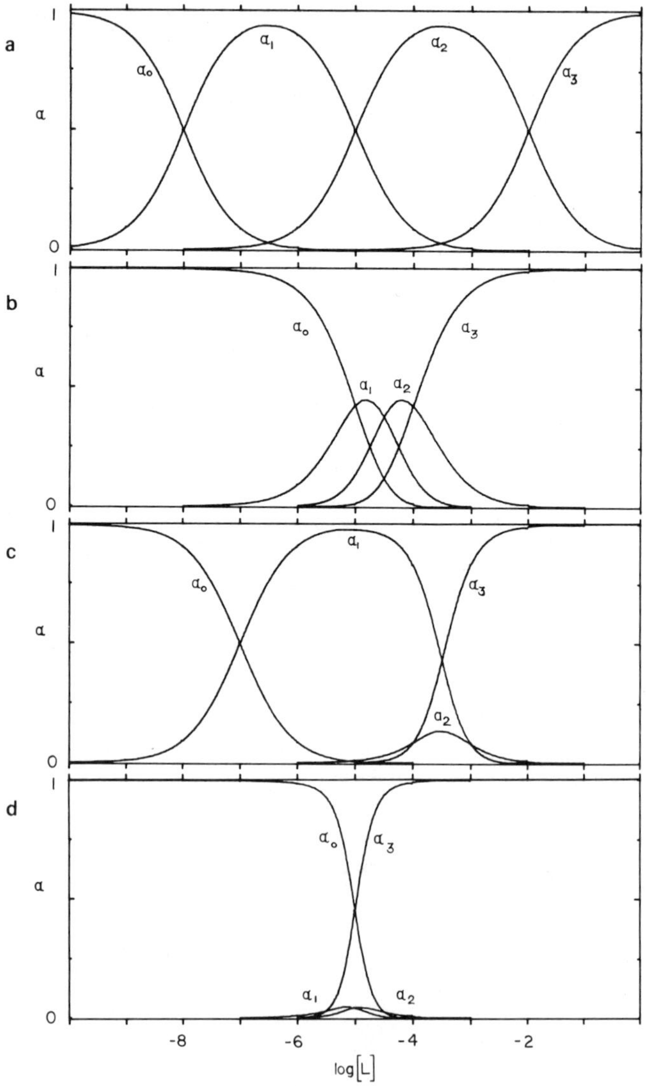

Figure 8-3. Four cases of three-constant systems; the α and $\bar{n}$ curves. See text for constant values.

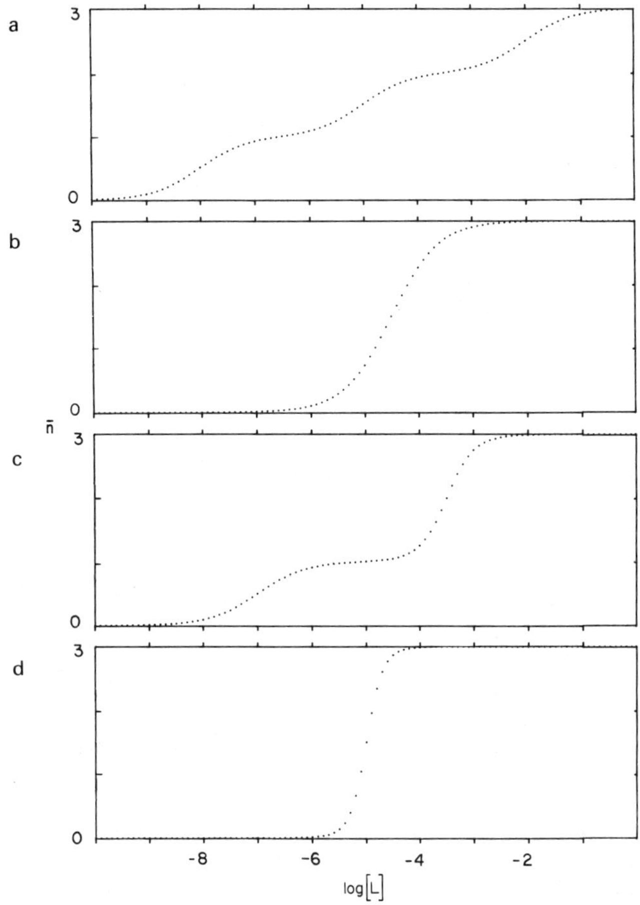

Figure 8-3 continued.

3. α_0 crosses α_2 at $\log[\mathrm{F}^-] = -\frac{1}{2}\log K_1K_2$. This is now only approximately the maximum position of α_1 (see $N = 2$ cases) because α_3 and the rest can affect this maximum, but not the $\log[\mathrm{F}^-]$ of the crossing. If α_3 and higher are near zero, then $\alpha_0 + \alpha_1 + \alpha_2 = 1$. Since $\alpha_0 = \alpha_2$ at this crossing,

$$\alpha_1 = 1 - 2\alpha_0 = [1 + 2(K_2/K_1)^{1/2}]^{-1}$$

4. The $\bar{n}$ (formation) curve reveals a method of finding K_n values when an experimental $\bar{n}$ curve can be obtained. Note that it goes approximately through the $-\log K_n$ values of $\log[\mathrm{F}^-]$ at half-

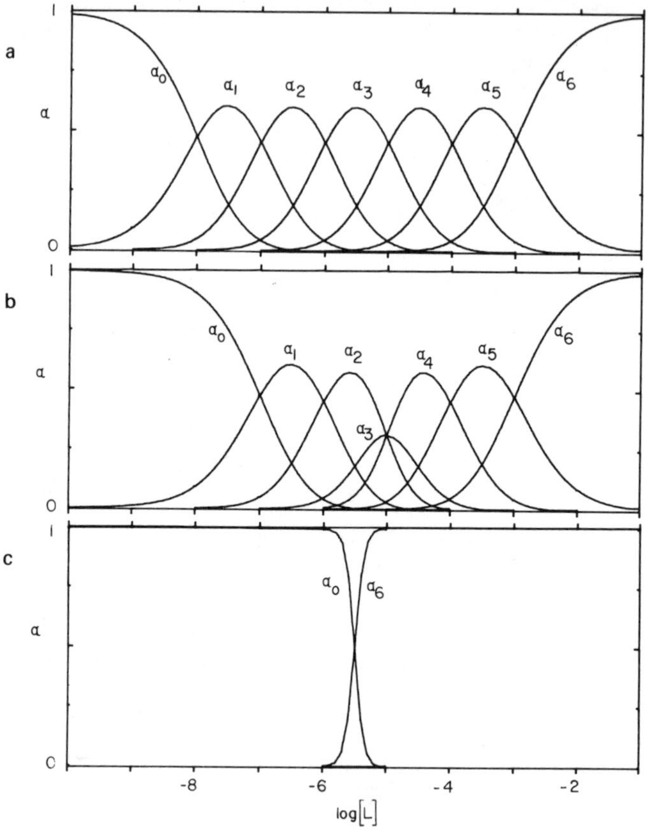

Figure 8-4. Three cases of six-constant systems; the α and $\bar{n}$ curves. See text for constant values.

integral $\bar{n}$ values:

$\bar{n}$	0.5	1.5	2.5	3.5	4.5	5.5
$\log[F^-]$	−6.16	−5.05	−3.91	−2.71	−1.46	0

This is the case due to the crossing situation already discussed: When α_0 crosses α_1, $\bar{n}$ must be 0.5 if the system contains mostly Al^{3+} and AlF^{2+} in equal amounts. Similarly, when the system is mainly AlF^{2+} and AlF_2^+ in equal amounts, $\bar{n} = 1.5$, and so on. (Equal amounts of Al^{3+} and AlF_3 will not shift this $\bar{n}$ from 1.5.) See the α curves in Figure 8-5. This has been a most important relationship in determination of equilibrium constants in solution.

Some interpretations and uses of these diagrams will be given now with numerical examples. Figure 8-5 tells at a glance that aquo

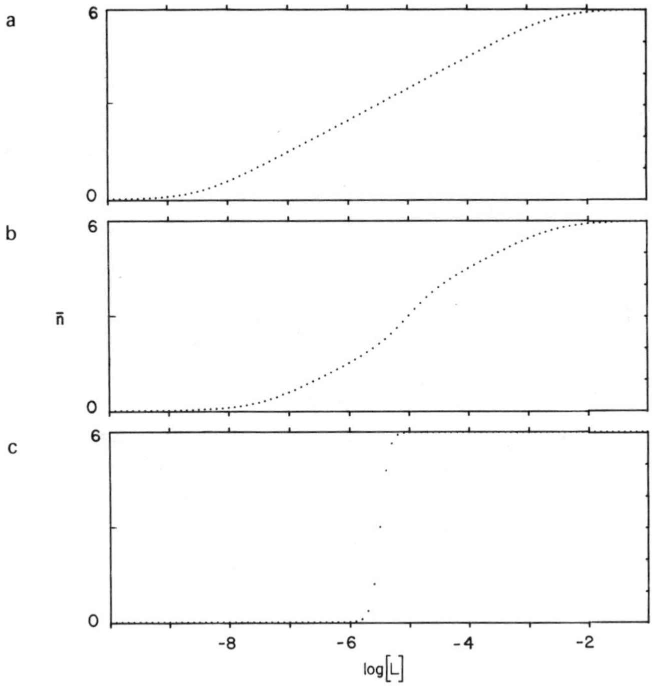

Figure 8-4 continued.

Al^{3+} ion is a minor species in any solution having uncomplexed $[F^-]$ above 10^{-5} M: that is, Al(III) is well complexed by F^- (masked). It tells us that three or more Al(III) species are present in appreciable proportions at all $[F^-]$ between 10^{-6} and 1 M. This is reflected in the $\bar{n}$ curve, which shows no abrupt changes or plateaus.

Attempts to describe any of these Al(III)–F^- solutions by use of one- or two-species approximations can lead to great error in interpretation of various measurements on them (spectra, cell voltages, etc.).

At$[F^-] = 10^{-3}$, AlF_3, AlF_4^-, AlF_2^+, and AlF_5^{2-} are present (in decreasing order) while AlF^{2+}, AlF_6^{3-}, and Al^{3+} are at much lower values of α. Note that very low values of α do not mean that the species can be neglected for all purposes. Minor species can be quite important in reaction kinetics and biochemical processes, for example.

When magnitudes of species concentrations are required, they can be obtained from the diagram, or equations (8-2) and (8-3), since

$$[ML_n] = \alpha_n C_M$$

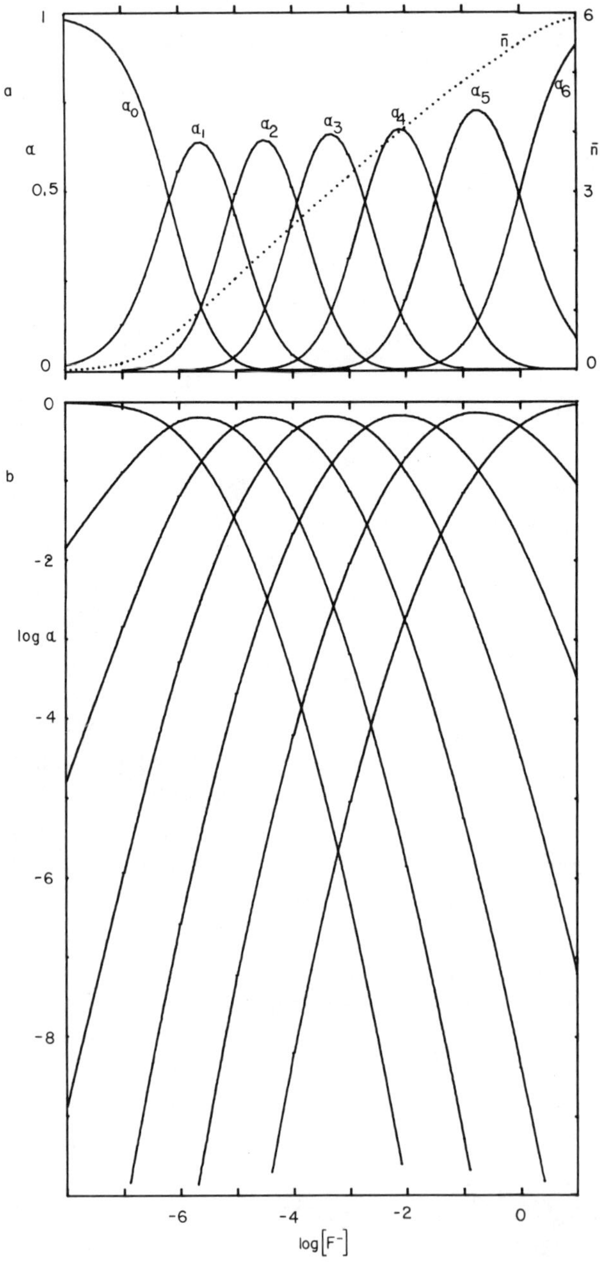

Figure 8-5. The system Al(III)–F^-; α, $\bar{n}$, and $\log \alpha$ curves.

Example. If KF is added to 0.0200 M (C_M) $Al(NO_3)_3$ solution until the uncomplexed $[F^-]$ becomes 10^{-3} M, we can use Figure 8-5a and/or equation (8-7) to obtain the required α values to use:

$$[Al^{3+}] = \alpha_0 C_M = 4.5 \times 10^{-7}(0.02) = 9.0 \times 10^{-9}\ M$$

$$[AlF^{2+}] = \alpha_1 C_M = 6.5 \times 10^{-4}(0.02) = 1.3 \times 10^{-5}\ M$$

$$[AlF_2^+] = \alpha_2 C_M = 0.074(0.02) = 1.48 \times 10^{-3}\ M$$

$$[AlF_3] = \alpha_3 C_M = 0.606(0.02) = 0.0121\ M$$

$$[AlF_4^-] = \alpha_4 C_M = 0.310(0.02) = 0.0062\ M$$

$$[AlF_5^{2-}] = \alpha_5 C_M = 0.009(0.02) = 1.8 \times 10^{-4}\ M$$

$$[AlF_6^{3-}] = \alpha_6 C_M = 9 \times 10^{-6}(0.02) = 1.8 \times 10^{-7}\ M$$

At this point, $\bar{n} = 3.25$ by equation (8-5). In this example we have assumed a **pH** about 4 where other complexing to form $AlOH^{2+}$ or HF can be taken to be small to a fair approximation. We shall have to treat such complications properly in many cases later.

To display the very small α values, log α plots are useful. Such a plot for the Al(III)–F^- system is shown in Figure 8-5. Values below 10^{-2} cannot be seen on the α plot, while values down to 10^{-10} are shown in the log plot.

4. Approximate Sketching of an Alpha Diagram

Quite rough α sketches can give valuable information about the species present and their relative amounts under various conditions. A summary of the regularities and fixed points we have observed in the foregoing cases should point the way to rapid construction of diagrams:

1. Choose coordinates: ordinate, α from 0 to 1; abscissa, log [L] from two units below $-\log K_1$ to two units above $-\log K_N$ (or use the corresponding **pH** and pK in acid–base systems).

2. Plot $\alpha_0 = 0.9$ and $\alpha_1 = 0.09$ at $\log[L] = -\log K_1 - 1$ and $\alpha_0 = 0.99$ and $\alpha_1 = 0.01$ at $\log[L] = -\log K_1 - 2$.

3. Locate the $-\log[L] = -\log K_n$ intersections of the adjacent α_n curves and place them according to spacings of the K_n values by observing the possible cases in Figure 8-2. Fill in the α curves with maxima half-way between $\log K_n = -\log[L]$ intersections.

4. Finish the highest two curves for α_N and α_{N-1} in symmetry with the first two in step 1 above.

After some practice, students can sketch these diagrams quite rapidly for use in approximate problem solving. Remember, the intersection log [L] values are quite correct, only the heights of the α curves are being approximated. The sum of all the alphas should equal 1. Thus, only one curve can be above 0.5 at any point.

5. Logarithmic Ratio Plots

These easily constructed diagrams consist entirely of straight lines. The mathematics was given in Chapter 5 for polyprotic acid–bases. We may choose the ratio of each species to any one of the others as a common basis. Let us illustrate with the Al(III)–F^- system. First with the ratios to Al(III) aquo ion, we use the β expressions directly, from set (5-2). Taking logs gives the convenient plotting form:

$$\beta_1 = 10^{6.164} = \frac{[ML]}{[M][L]}, \qquad \log\frac{[ML]}{[M]} = 6.164 + \log[L]$$

$$\beta_2 = 10^{11.217} = \frac{[ML_2]}{[M][L]^2}, \qquad \log\frac{[ML_2]}{[M]} = 11.217 + 2\log[L]$$

$$\beta_3 = 10^{15.13} = \frac{[ML_3]}{[M][L]^3}, \qquad \log\frac{[ML_3]}{[M]} = 15.13 + 3\log[L]$$

$$\beta_4 = 10^{17.84} = \frac{[ML_4]}{[M][L]^4}, \qquad \log\frac{[ML_4]}{[M]} = 17.84 + 4\log[L]$$

$$\beta_5 = 10^{19.30} = \frac{[ML_5]}{[M][L]^5}, \qquad \log\frac{[ML_5]}{[M]} = 19.30 + 5\log[L]$$

$$\beta_6 = 10^{19.30} = \frac{[ML_6]}{[M][L]^6}, \qquad \log\frac{[ML_6]}{[M]} = 19.30 + 6\log[L]$$

The log equations give lines of slopes 1–6 and intercepts at $\log\beta$, $\log[L] = 0$ (1 M F^-) for plots of $\log R$ vs. $\log[L]$. We obtain a diagram that is somewhat easier to read if we take one of the central species as the reference for our ratios. It is a simple matter to convert our set of equations above to the new ratios. Using AlF_3 as the reference, we divide each β_n by β_3 (subtract the corresponding log equations). We use $R_{3,3} = 1$, constant line for reference on the graph (Figure 8-6). The new set of equations is

$$R_{0,3} = \frac{[M]}{[ML_3]} = (\beta_3[L]^3)^{-1}, \quad \log R_{0,3} = -15.13 - 3 \log [L]$$

$$R_{1,3} = \frac{[ML]}{[ML_3]} = \frac{\beta_1}{(\beta_3[L]^2)}, \quad \log R_{1,3} = -8.96 - 2 \log [L]$$

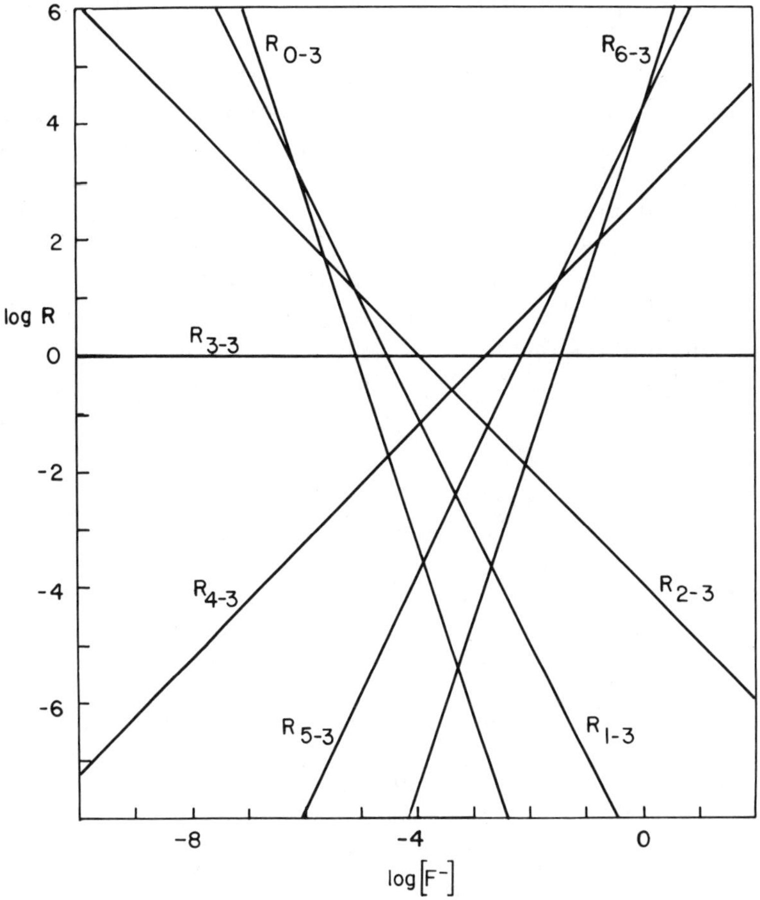

Figure 8-6. The log ratio diagram for the system Al(III)–F⁻.

$$R_{2,3} = \frac{[ML_2]}{[ML_3]} = \frac{\beta_2}{(\beta_3[L])}, \qquad \log R_{2,3} = -3.91 - \log [L]$$

$$R_{3,3} = \frac{[ML_3]}{[ML_3]} = 1, \qquad \log R_{3,3} = 0$$

$$R_{4,3} = \frac{[ML_4]}{[ML_3]} = \frac{\beta_4[L]}{\beta_3}, \qquad \log R_{4,3} = 2.71 + \log [L]$$

$$R_{5,3} = \frac{[ML_5]}{[ML_3]} = \frac{\beta_5[L]^2}{\beta_3}, \qquad \log R_{5,3} = 4.17 + 2 \log [L]$$

$$R_{6,3} = \frac{[ML_6]}{[ML_3]} = \frac{\beta_6[L]^3}{\beta_3}, \qquad \log R_{6,3} = 4.17 + 3 \log [L]$$

This conversion may impress the reader with the useful point that any species in step equilibrium can be expressed as a ratio to any other in terms of the constants and only one other variable, commonly log [L] or the **pH**.

Although the maxima are not clearly visible, these diagrams can serve many purposes of the α diagrams. For example, a trace of Al(III) in a natural water containing 10^{-4} M F^- should be about half AlF_3 and half AlF_2^+. The relative amounts of the other species can be read off as fractions of the AlF_3 as follows: Al^{3+} 10^{-3}, AlF^{2+} 10^{-1}, AlF_4^- 0.05, AlF_5^{2-} 10^{-4}, AlF_6^{3-} $10^{-7.5}$. For non-trace quantities,

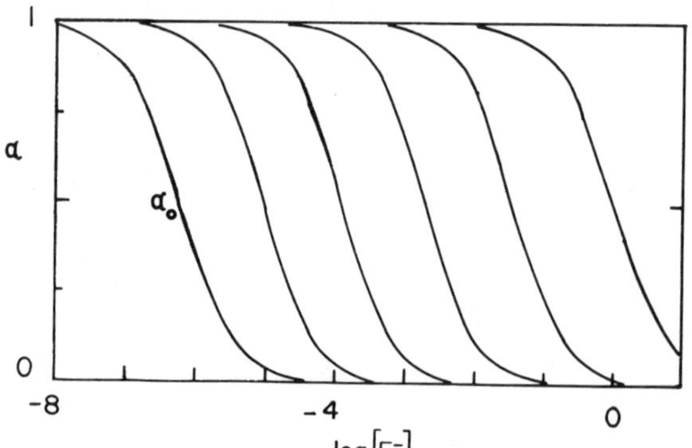

Figure 8-7. The abundance area diagram for the system Al(III)–F^-.

where C_M and C_L are not the equilibrium values, the $\bar{n}$ intersection method can be used.

One further plotting method is shown in Figure 8-7. It is an abundance area diagram, in which the α value of each species is plotted as an added line segment up from the previous ones. The first line at the left is simply α_0. The second line is $\alpha_0 + \alpha_1$, the third $\alpha_0 + \alpha_1 + \alpha_2$, and so on to $\alpha_0 + \alpha_1 + \alpha_2 + \alpha_3 + \alpha_4 + \alpha_5$. Thus, the vertical distance between two lines is the α value of the species in that area. α_6 is the distance from the last line to the top. These diagrams offer no advantages over the others presented, and do not show maxima clearly. They will not be used further in this book.

Selected Reading[4]

DAVIES, C. W., *Ion Association*, Butterworths, London, 1962.

JONES, MARK M., *Elementary Coordination Chemistry*, Prentice-Hall, Englewood Cliffs, New Jersey, 1964.

BASOLO, F., and PEARSON, R. G., *Mechanisms of Inorganic Reactions*, Wiley, New York, 1967, Chapters 1 and 2.

Problems

1. Sketch roughly the α and $\bar{n}$ diagrams for the following systems having four log K values as given. Label the α_n's. (a) 8, 6, 4, 2. (b) 5, 4, 3, 2. (c) 5, 3, 4, 4. (d) 4, 4, 4, 4.

2. Differentiate α_1^{-1} to prove the expression given in this chapter for its maximum for two-K cases. Calculate its numerical value for the cases in Figure 8-2.

3. For the unsymmetrical case for three constants (log K of 7, 3, and 4) in Figure 8-3c, does the maximum of α_1 occur over the crossing of α_0 and α_2? Give a proof.

4. Estimate log [L] by plotting $\bar{n}'$ roughly on the $\bar{n}$ diagram of Figure 8-3 for each case when the solution is made to have analytical concentration 0.0100 M ML_2.

5. Estimate the concentrations of all the Al(III) species in a solution of analytical concentrations, Al(III) 0.0500 M and F^- 0.1000 M. Use the various diagrams of this system in this chapter.

---

[4]Also see the list in Chapter 5.

9 Applications of Metal Ion–Ligand Calculations

The α diagrams of Figures 9-1–9-3 point up some relationships among elements with respect to the periodic table and their electronic structures.

Copper(I) is smaller and harder as a Lewis acid than Ag(I), both d^{10} ions. Copper(I) is more tightly bound by the hard base, NH_3. The low charge favors coordination of only two ligands at open *s* and *p* orbitals in a linear configuration. The $+2$ charge of the d^9 Cu(II) helps increase the coordination to four rather strongly held ammonias and two additional ones more weakly bound. The d^{10} ions Zn(II), Cd(II), and Hg(II) show predominantly tetrahedral complexing. Mercury(II) seems unique with its characteristically spaced pairs of formation constants: two ligands strongly held in the linear form and two ligands weakly held in orbitals reorganized to tetrahedral geometry. Note the steady increase in stability of the Hg(II) complexes with increasing softness of the ligand: Cl^-, Br^-, I^-, CN^-.

Open *d* orbitals are available in Mg(II), d^0; Co(II), d^7; and Ni(II), d^8; which exhibit octahedral, 6-coordination. In all cases, it is important to remember that we deal with high concentrations of the competing ligand H_2O which fills all the positions not mentioned in our conventional abbreviated way of writing these formulas. Thus $HgCl^+$ means $HgCl(H_2O)_3^+$, and $Cu(NH_3)_4^{2+}$ means $Cu(NH_3)_4(H_2O)_2^{2+}$.

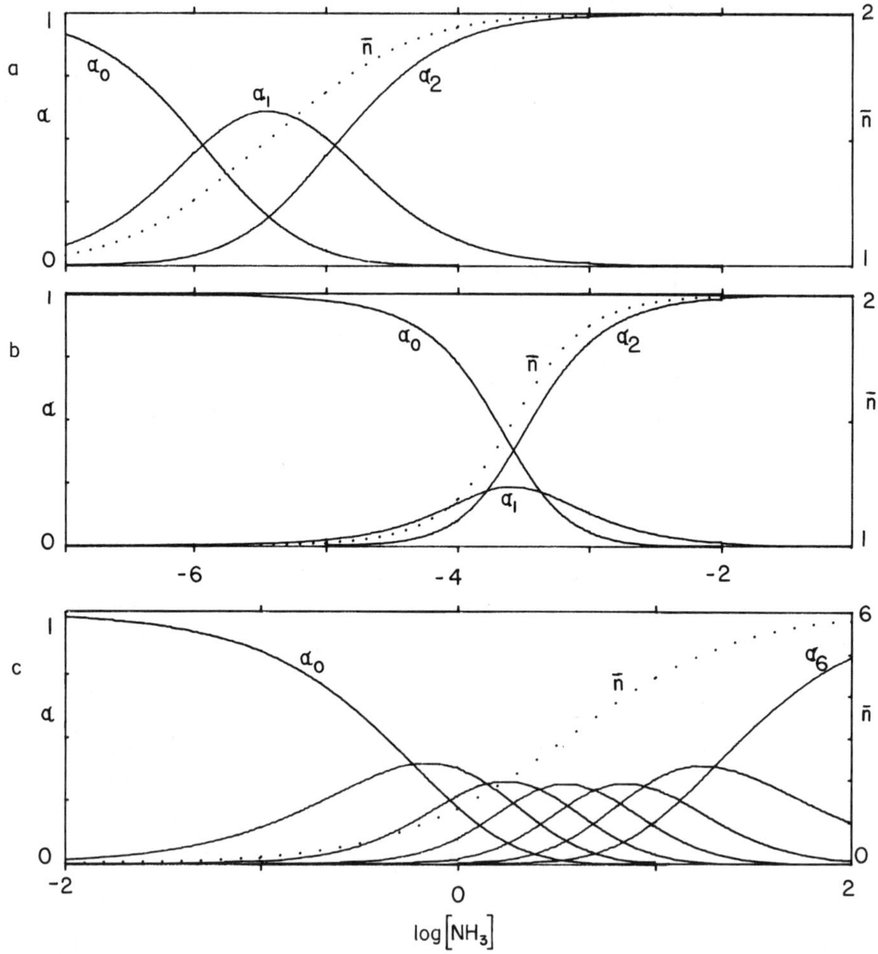

Figure 9-1. Metal ion–ammonia systems. See Table 9-1 for conditions and constants, α curves solid, $\bar{n}$ dotted. (a) Cu(I), (b) Ag(I), (c) Mg(II), (d) Co(II), (e) Ni(II), (f) Cu(II), (g) Zn(II), (h) Cd(II), (i) Hg(II).

The constants used in these computer plots are listed in Table 9-1.[1] Further discussion and examples are available in the references in Chapter 8. Linear relations among the formation constants of a large number of hard and soft ligands when plotted as free

[1]From L. G. Sillen and A. E. Martell, *Stability Constants of Metal-Ion Complexes*, Special Publication No. 17, 1964, and No. 25, 1971, The Chemical Society, London.

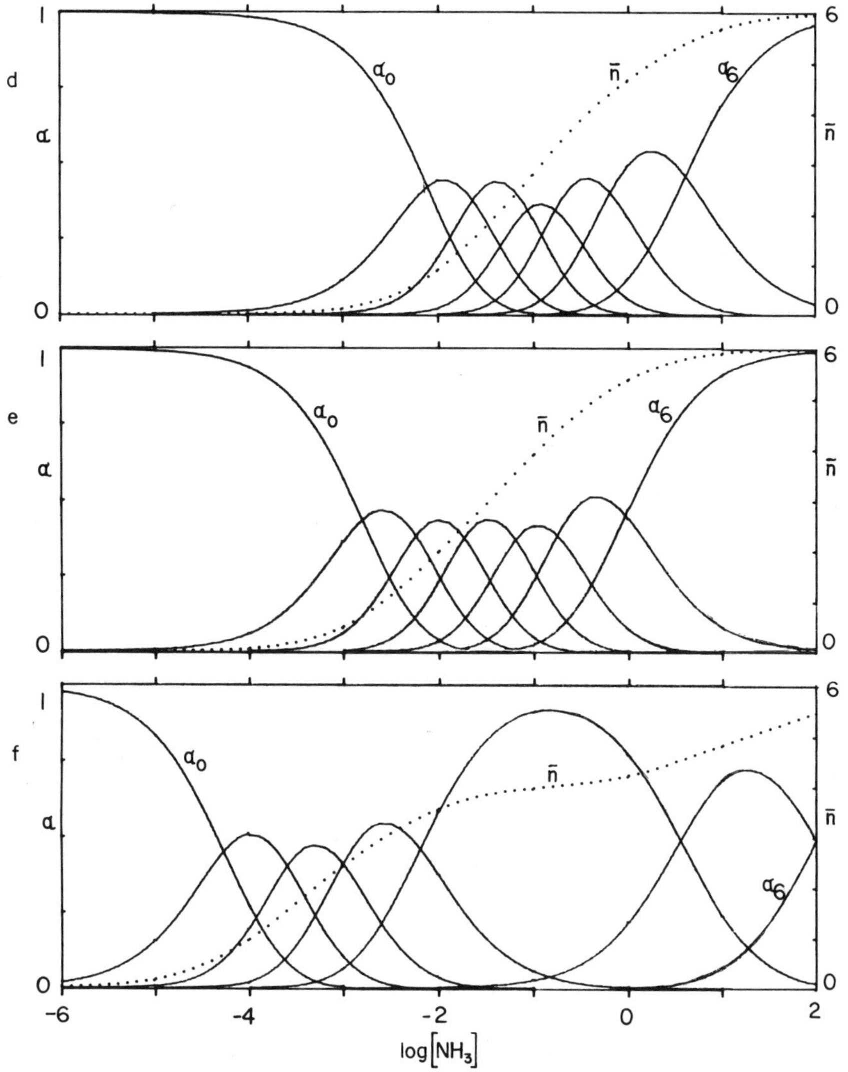

Figure 9-1d, e, f

energy values are demonstrated by Hancock, Finkelstein, and Evers.[2]

Let us examine a few well-studied systems. Further examples adding precipitation and redox reactions will be treated later in Chapters 11 and 12.

[2]R. D. HANCOCK, N. P. FINKELSTEIN, and A. EVERS, *J. Inorg. Nucl. Chem.*, **36**, 2539 (1974).

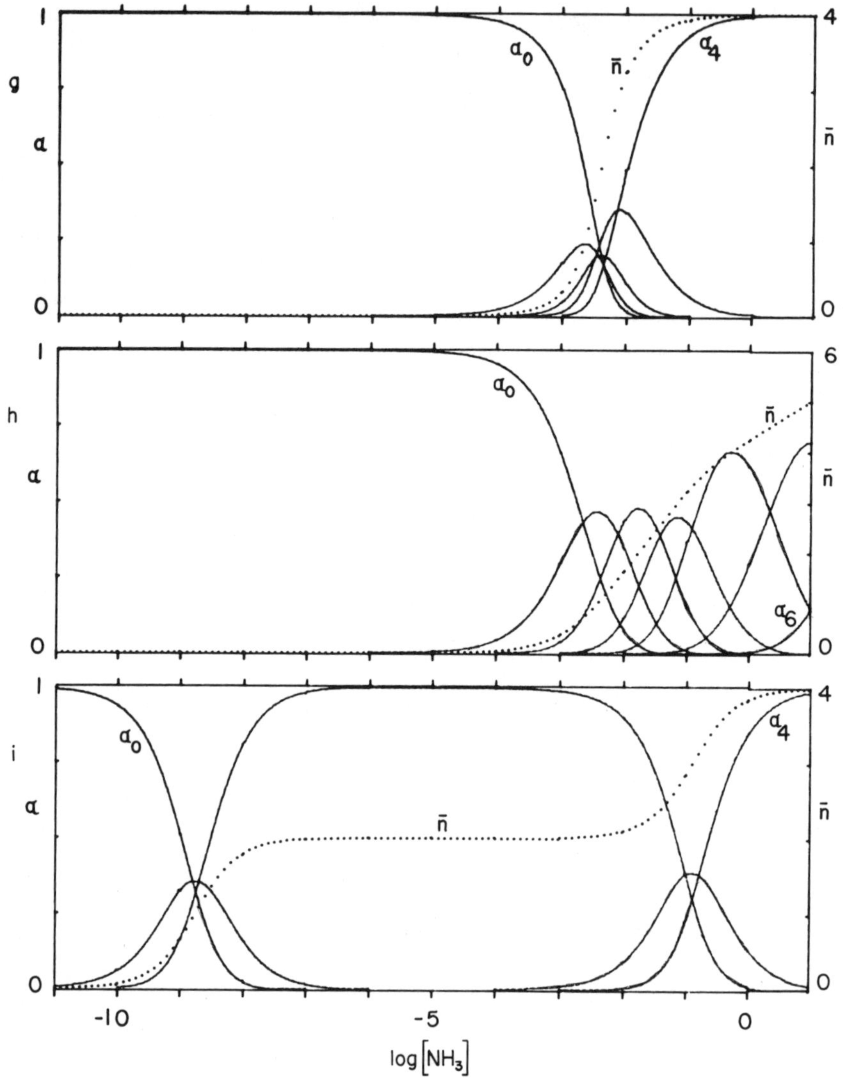

Figure 9-1g, h, i

1. Fe(III)–SCN^-

Figure 9-4 shows the α and $\bar{n}$ diagrams for this system at 25° in a medium of high constant ionic strength, 1.2 M $NaClO_4$. The intensely red complexes of iron(III) with thiocyanate have long been used for detection and determination of iron. The log K_n values used here are

Table 9-1. Constants and Conditions for Figures 9-1–9-3

Metal ion	T, °C	Medium	log formation constant					
			K_1	K_2	K_3	K_4	K_5	K_6
Ammonia complexes								
Cu(I)	18	2 M NH_4NO_3	5.93	4.93	—	—	—	—
Ag(I)	25	1 M NH_4NO_3	3.37	3.78	—	—	—	—
Mg(II)	23	2 M NH_4NO_3	0.23	−0.15	−0.42	−0.7	−0.95	−1.3
Co(II)	30	2 M NH_4NO_3	2.11	1.63	1.05	0.76	0.18	−0.62
Cu(II)	25	2 M NH_4NO_3	4.27	3.59	3.00	2.19	−0.55	−2.0
Ni(II)	30	2 M NH_4NO_3	2.80	2.24	1.73	1.19	0.75	0.03
Zn(II)	30	2 M NH_4NO_3	2.37	2.44	2.50	2.15	—	—
Cd(II)	30	2 M NH_4NO_3	2.65	2.10	1.44	0.93	−0.32	−1.66
Hg(II)	23	2 M NH_4NO_3	8.8	8.7	1.00	0.78	—	—
Halides, cyanides								
Zn(II)–Cl^-	20	0.69 M $HClO_4$	0.72	−0.23	−0.68	+0.37	—	—
Cd(II)–Cl^-	20	2.1 M KNO_3	1.77	1.45	−0.25	−0.05	—	—
Cd(II)–Br^-	25	1 M $NaClO_4$	1.56	0.46	0.23	0.41	—	—
Cd(II)–I^-	25	0.25 M $NaClO_4$	1.94	0.70	1.68	1.19	—	—
Hg(II)–Cl^-	25	0.5 M $NaClO_4$	6.74	6.48	0.95	1.05	—	—
Hg(II)–Br^-	25	0.5 M $NaClO_4$	8.94	7.94	2.27	1.75	—	—
Hg(II)–I^-	25	0.5 M $NaClO_4$	12.87	10.95	3.67	2.37	—	—
Hg(II)–CN^-	20	0.1 M $NaNO_3$	18.00	16.70	3.83	2.98	—	—

2.11, 1.19, 0, 0, −0.1, −0.2. Only the first two constants are well determined, due to experimental problems. The methods of Chapters 5 and 8 are applicable to calculation of species concentrations. When log [SCN^-] can be estimated at equilibrium, it is a simple matter to read off the α values. However, let us carry out the more usual requirement of finding these from initial analytical concentrations.

Example 1. What are the major species and their concentrations after mixing equal volumes of 0.20 M Fe^{3+} and SCN^- solutions? Assume 1.2 ionic strength. C_M and C_L are both 0.10 M after mixing. Following the example of equation (8-4), we have

$$\bar{n}' = (C_L - [L])/C_M = (0.1 - [SCN^-])/0.1 = 1 - [SCN^-]/0.1$$

Choose several values of [SCN^-] and calculate $\bar{n}'$. Plot these on the equilibrium $\bar{n}$ line graph to find the intersection, the solution to the

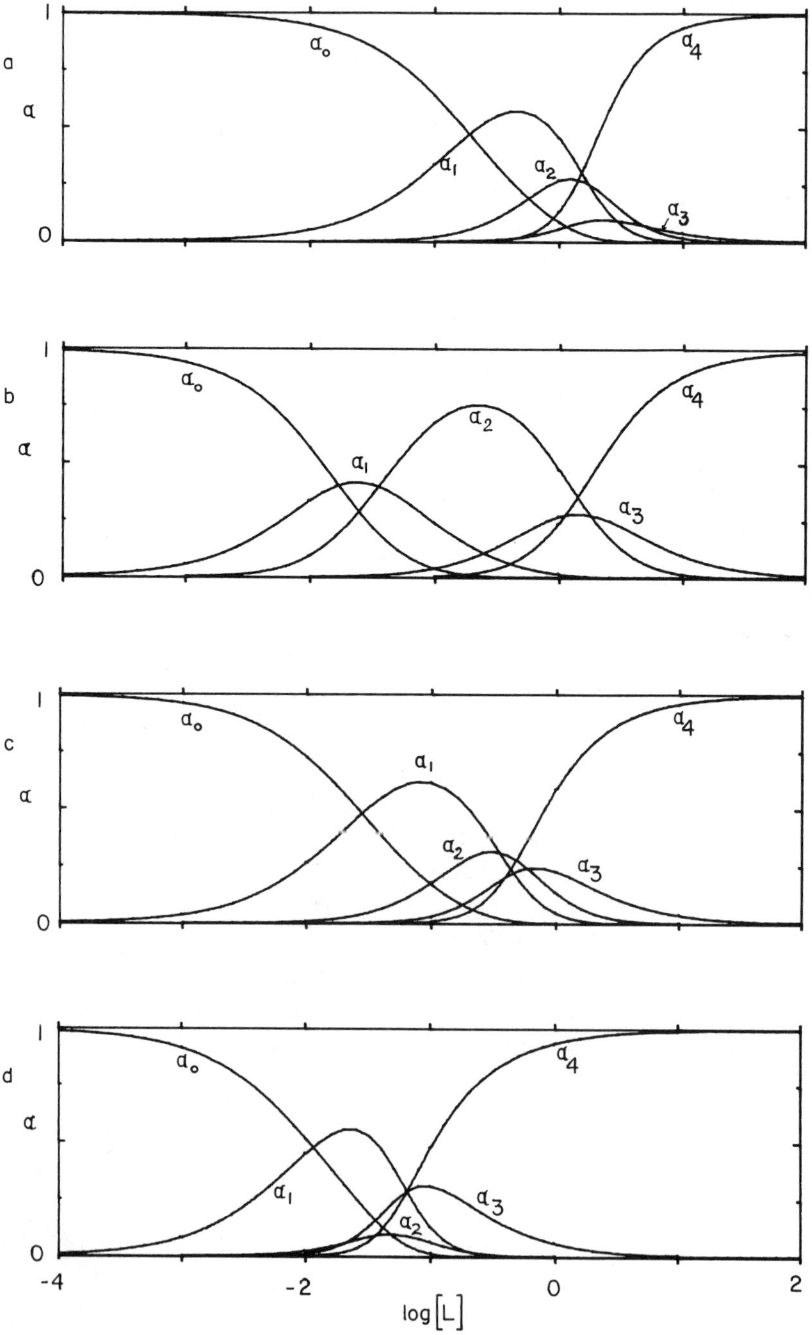
a
b
c
d
α
1
0
α₀
α₁
α₂
α₃
α₄
-4
-2
0
2
log[L]

Figure 9-2. Metal ion–halide and cyanide systems. See Table 9-1 for conditions and constants. α curves. (a) Zn(II)–Cl^-, (b) Cd(II)–Cl^-, (c) Cd(II)–Br^-, (d) Cd(II)–I^-, (e) Hg(II)–Cl^-, (f) Hg(II)–Br^-, (g) Hg(II)–I^-, (h) Hg(II)–CN^-.

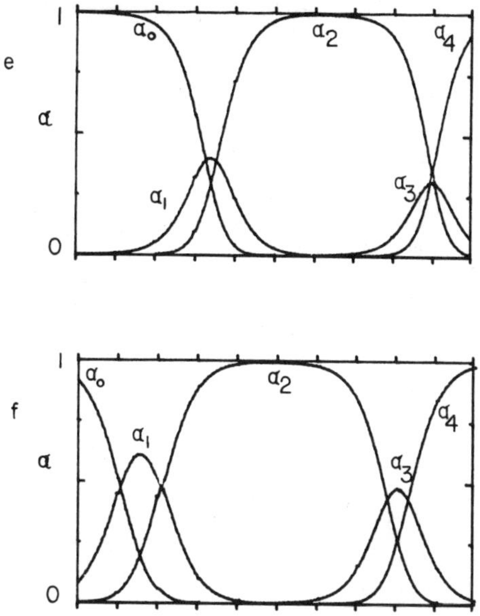

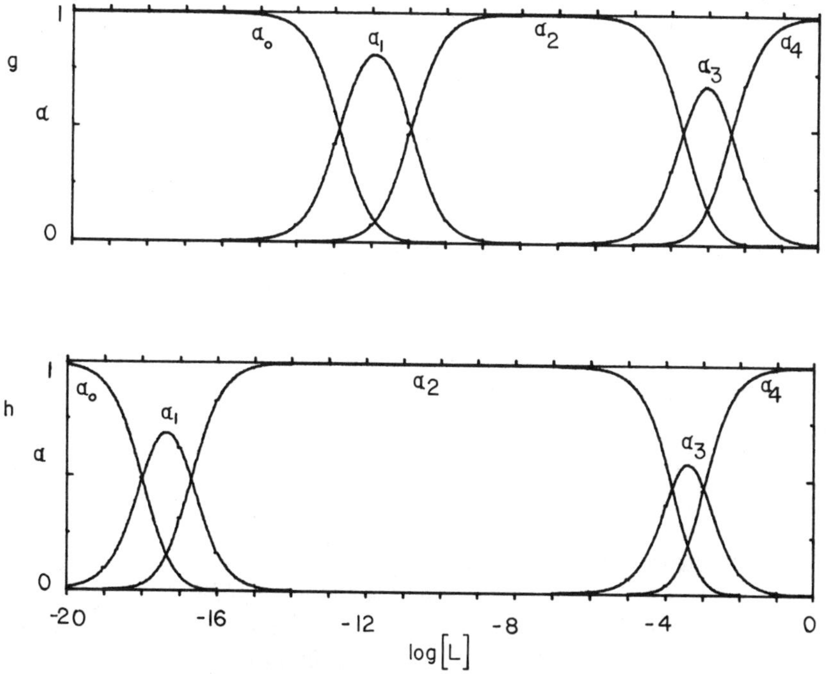

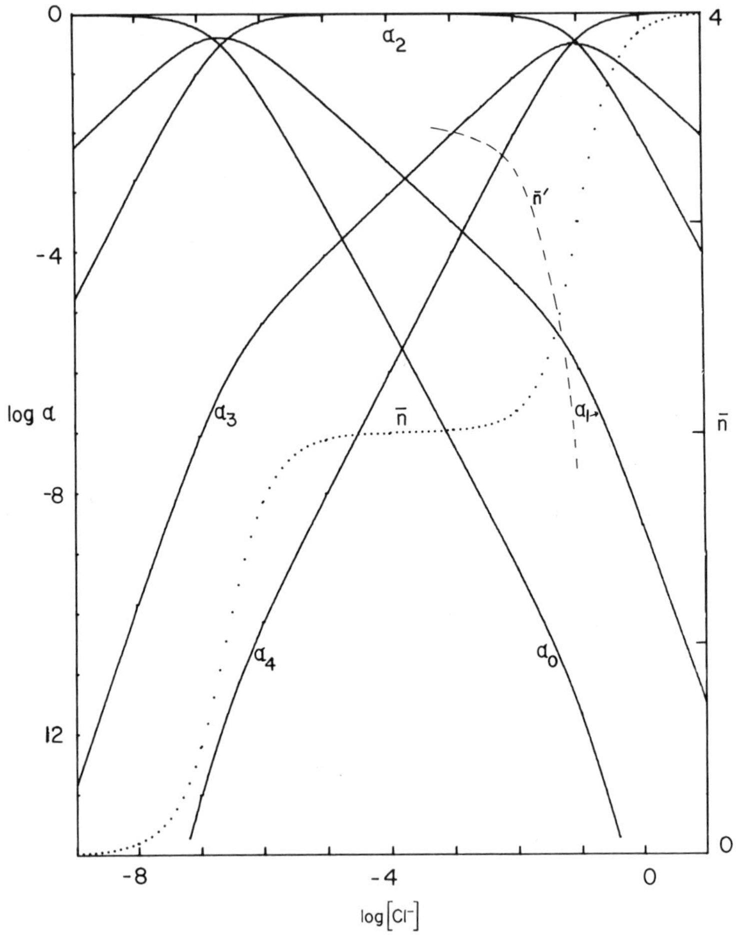

Figure 9-3. Log α and $\bar{n}$ for the system Hg(II)–Cl^-. The dashed line for $\bar{n}'$ is for the solution of Example 2.

simultaneous equations for this system.

$[SCN^-]$	0.05	0.02	0.01	0.015
$\bar{n}'$	0.5	0.80	0.90	0.85
log [L]	−1.3	−1.7	−2	−1.82

The intersection found graphically is at $\bar{n} = 0.85$, $\log [SCN^-] = -1.82$. At this log [L] on the α diagram, one reads $\alpha_0 = 0.28$, $\alpha_1 = 0.57$, and $\alpha_2 = 0.14$ for major species of Fe(III). These, times 0.10, give concentrations of each: Fe^{3+}, $FeSCN^{2+}$, and $Fe(SCN)_2^+$.

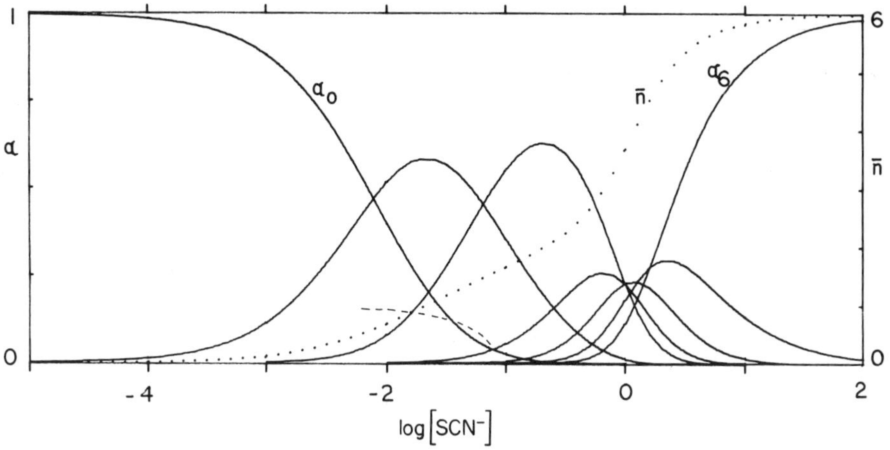

Figure 9-4. α and $\bar{n}$ curves for the system Fe(III)–SCN^-. Dashed $\bar{n}'$ line for Example 1 solution.

Since this is the [SCN^-] region in which colorimetric analysis is done, one may ask how reliable results can be obtained with such a distribution of species. The Fe^{3+} is of negligible color compared to the dark red of the thiocyanate complexes. Only if the same unbound [SCN^-] is present in both standard and unknown solutions can the method be reliable. Then the ratio of iron(III) in the two is correct due to cancelling of the α's. For example, if a Dubosq colorimeter is used to find a ratio of lengths of solutions to give equal color intensity, we have, since length is inversely proportional to apparent color,

$$\frac{\text{red iron in standard}}{\text{red iron in unknown}} = \frac{(\alpha_1 + \alpha_2)C_{\mathrm{Fe}(s)}}{(\alpha_1 + \alpha_2)C_{\mathrm{Fe}(u)}} = \frac{l(u)}{l(s)}$$

Thus, if [SCN^-] is made the same in both solutions, the α values are the same and the lengths are inversely related to total iron in the solutions.

2. Hg(II)–Cl^-

The α diagram for the mercury(II)–chloride system is given in Figure 9-2, and the log α and $\bar{n}$ curves in Figure 9-3. Note the unusual inverted order of the K_3 and K_4, which means that $HgCl_3^-$ is never the species of highest α. The system can be treated well in two separate parts, one like Figure 8-2c and the other like Figure 8-2e. Calculations can be based on the following approximations: in solutions of $[Cl^-]$ above 10^{-3} M, $HgCl_2$, $HgCl_3^-$, and $HgCl_4^{2-}$ are practically the only mercury species present, and below 10^{-5} M $[Cl^-]$, Hg^{2+}, $HgCl^+$, and $HgCl_2$ are the mercury species.

Example 2. Find the composition of the solution made by mixing equal volumes of 0.100 M $HgCl_2$ and 0.160 M HCl. Here $C_L = 0.100 + 0.080 = 0.180M$, and $C_M = 0.050$ M As above, we find the $\bar{n}'$ line,

$$\bar{n}' = (0.180 - [Cl^-])/0.050 = 3.60 - 20[Cl^-]$$

This line is plotted on Figure 9-3 and found to intersect $\bar{n}$ at $\log[Cl^-] = -1.28$. At this value, we can read the α values from Figure 9-2 as $\alpha_2 = 0.64$, $\alpha_3 = 0.22$, and $\alpha_4 = 0.14$. For the smaller values we go to the $\log \alpha$ curves of Figure 9-3 to read $\alpha_0 = 10^{-11}$ and $\alpha_1 = 10^{-5.5}$. These can be multiplied by 0.050 M to obtain the concentrations of the mercury(II) species. If the $\log \alpha$ curves were not available, we could substitute the concentrations of Cl^- and $HgCl_2$ into K_1 and K_1K_2 expressions to get the smaller species concentrations.

3. Cu(II)–NH_3–EDTA: Competing Ligands

Copper(II) has often been colorimetrically determined by its intensely blue ammonia complexes. The α diagram in Figure 9-1 shows that a similar situation occurs to that for the Fe(III)–SCN^- complexes, namely that a constant excess NH_3 concentration will be required if the color is to be proportional to the Cu(II) concentration. Even more useful, now, is the use of EDTA in the determination of metal ions. Copper ion is often titrated in ammonia buffers. Let us examine the constants to see if quantitative complexing of Cu(II) by EDTA should occur in the presence of NH_3.

Example 3. What proportion of Cu(II) remains uncomplexed by EDTA at equivalence when the titration is performed in ammonia solution at **pH** 10 with $[NH_3]$ 0.2 M? Let the total copper(II) be 0.010 M at equivalence. The EDTA forms 1:1 complexes with metal ions and the K_1 with Cu^{2+} is $10^{18.8}$ (Table 9-2). The ligand in question is the Y^{4-} ion, whose α_0 curve is shown in Figure 9-5. At **pH** 10, α_{0L} is $10^{-0.45}$. We also must find the α_{0M} for the fraction of Cu(II) left uncomplexed by NH_3, and thus free to enter into the EDTA com-

Table 9-2. Formation Constants for EDTA Complexes[a]

Metal ion	$\log K_1$	Metal ion	$\log K_1$
Co^{3+}	36	Al^{3+}	16.1
Fe^{3+}	25.1	La^{3+}	15.5
Th^{4+}	23.2	Fe^{2+}	14.3
Cr^{3+}	23	Mn^{2+}	13.8
Bi^{3+}	22.8	Ca^{2+}	10.7
Hg^{2+}	21.8	Mg^{2+}	8.7
Cu^{2+}	18.8	Sr^{2+}	8.6
Ni^{2+}	18.6	Ba^{2+}	7.8
Pb^{2+}	18.0	Ag^{+}	7.3
Zn^{2+}	16.5	Li^{+}	2.8
Cd^{2+}	16.5	Na^{+}	1.7
Co^{2+}	16.3		

[a] Rounded values for ionic strength 0.1 M and temperature 20–25°C.

plexing system. We then use the formation expression for the copper–EDTA complex,

$$K_1 = 10^{18.8} = \frac{[CuY^{2-}]}{[Cu^{2+}][Y^{4-}]} = \frac{0.010 - x}{x\alpha_{0M}x\alpha_{0L}} \qquad (9\text{-}1)$$

This expression is obtained by assuming that most of the Cu(II) will be present as the EDTA complex, CuY^{2-}, approximately 0.01 M. The small amount x that does dissociate enters the Cu^{2+}–NH_3 system, and the Y^{4-} enters the H_4Y protonic system. Thus, we approach this problem by conceiving the process as placing 0.010 mol of CuY^{2-} in one liter of the 0.2 M ammonia buffer and allowing it to go to equilibrium.

What is α_{0M}? Figure 9-1 shows us that over 90% of the Cu(II) not in the EDTA complex will be $Cu(NH_3)_4^{2+}$. This makes it possible for us to take that tetraamine as the total Cu(II) in the ammonia system, to one significant figure, which gives an approximate α_{0M} as the ratio of $[Cu^{2+}]/[Cu(NH_3)_4^{2+}]$, which we can easily find from the β_4 expression:

$$\beta_4 = 10^{13.05} = \frac{[Cu(NH_3)_4^{2+}]}{[Cu^{2+}][NH_3]^4} = \frac{x}{[Cu^{2+}](0.2)^4}$$

$$\alpha_{0M} \cong \frac{[Cu^{2+}]}{x} = [10^{13.05}(0.2)^4]^{-1} = 10^{-10.25}$$

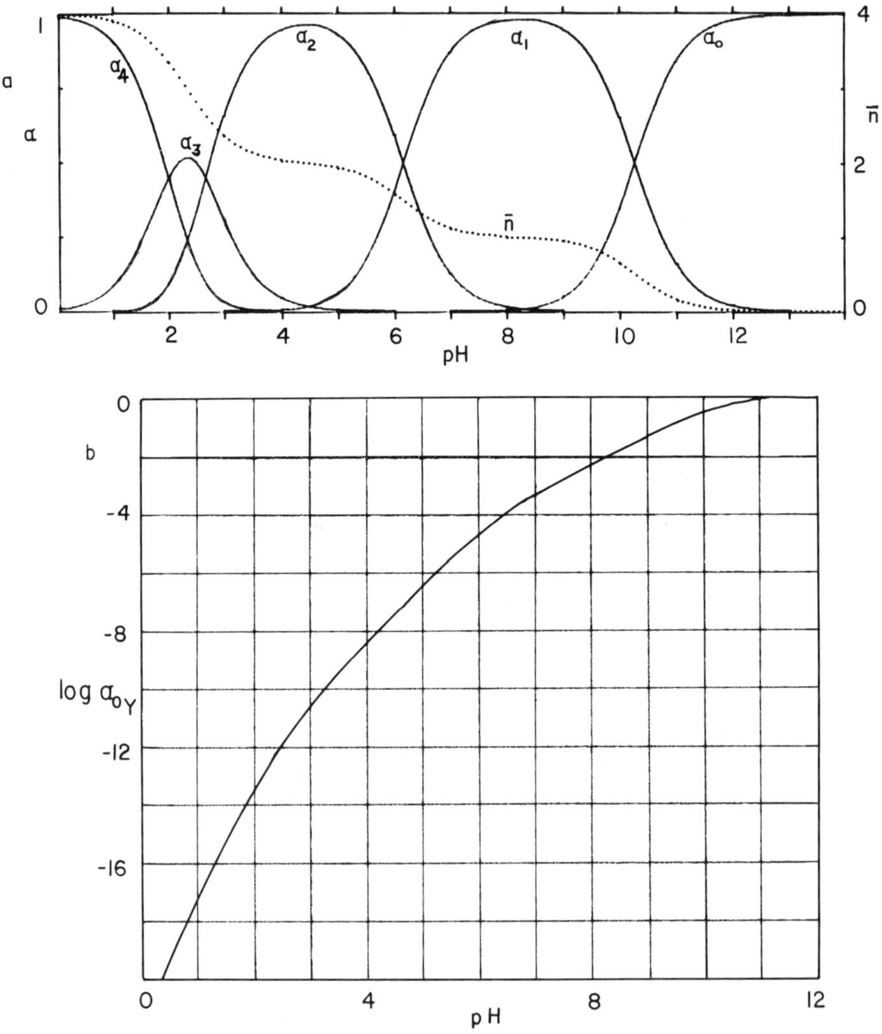

Figure 9-5. (a) The α and $\bar{n}$ curves for EDTA, 25°, 0.1 M ionic strength. (b) The log α_{0Y} curve for use in calculations involving the EDTA Y^{4-} ion complexing. (The highly protonated forms H_5Y^+ and H_6Y^{2+} have been omitted. These would appear below pH 2, where EDTA has low solubility. The pK_a's are about 1.4 and -0.1.)

This compares with $10^{-10.23}$ calculated from the complete α_0 equation. This is equivalent to taking the last term of equation (8-2) as the major one for α_0. Substitution of this value in equation (9-1) gives $x = 10^{-5.05}$. This is about 0.1 %, which fulfills our approximation and shows that the reaction is effectively quantitative at these conditions.

Table 9-3

pH	$\log \alpha_{0Y}$	$\log x$	$\alpha_{0Zn} = x/0.01$
3	−10.6	−3.95	0.011
2	−13.5	−2.57	0.27

4. The pH Effect in EDTA Complexing

The success of EDTA titration depends on the high K_f values for most metal ions (Table 9-2). Since the ligand is taken to be Y^{4-}, a rather strong base, the **pH** will be important in determining its effective concentration. The $[Y^{4-}]$ is given by α_{0Y} times the EDTA not in MY, the metal ion complex.

Let us examine the **pH** effect on complexing of 0.01 *M* ZnY^{2-}. Assume that no other ligand is present for Zn^{2+}, $\alpha_{0M} = 1$. As in the previous example, we use the K_f expression to get

$$K_f = 10^{16.50} = \frac{[ZnY^{2-}]}{[Zn^{2+}][Y^{4-}]} = \frac{0.01 - x}{x\alpha_{0Y}x}$$

Here, x is the part of the complex that dissociates at equilibrium. The **pH** dependence is contained in the α_{0Y} term (Figure 9-5). We can calculate x, and from it, the fraction of complex formed (Table 9-3). Figure 9-6 is plotted in this way. One can see the reason that EDTA titration of Zn^{2+} is directed at **pH** above 5, where the reaction is quantitative.

We can now get the minimum pH condition for quantitative reaction with EDTA (99.9%) using the K_f expression. With 0.01 *M* final metal ion, one wants only 10^{-5} *M* uncomplexed (0.1%). At the

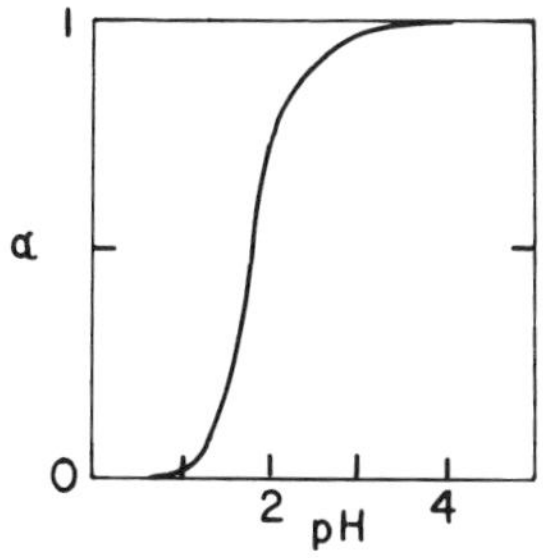

Figure 9-6. Fraction of Zn(II) in EDTA complex at various **pH**.

Table 9-4

Ion	$\log K_f$	$\log \alpha_{0Y}$ (min)	pH (min)
Ca^{2+}	10.70	−2.70	7.6
Zn^{2+}	16.50	−8.50	4.0
Hg^{2+}	21.80	−13.80	1.9
Fe^{3+}	25.1	−17.1	1.0

equivalence point [as in equation (9-1)],

$$K_f = \frac{[MY]}{x \cdot x\alpha_{0Y}} = \frac{0.01}{10^{-5} \cdot 10^{-5}\alpha_{0Y}}$$
$$\alpha_{0Y} = (K_f \cdot 10^{-8})^{-1} \qquad (9\text{-}2)$$

For four important cases this yields the results given in Table 9-4. These results agree with stipulations in laboratory methods. A plot of $\log K_f$ for EDTA complexes vs. this minimum pH is given in Figure 9-7. This shows that the lowest $\log K_f$ that can be used is about 8, the value for Mg^{2+}, approximately. This requires pH about 11. Most of the transition metal ions have $\log K_f$ above 14 for EDTA and can be titrated in acidic solutions. This has advantages in the choices of indicators it makes possible.

Masking, competition of other ligands for the metal ion in EDTA methods, can be treated by similar steps.

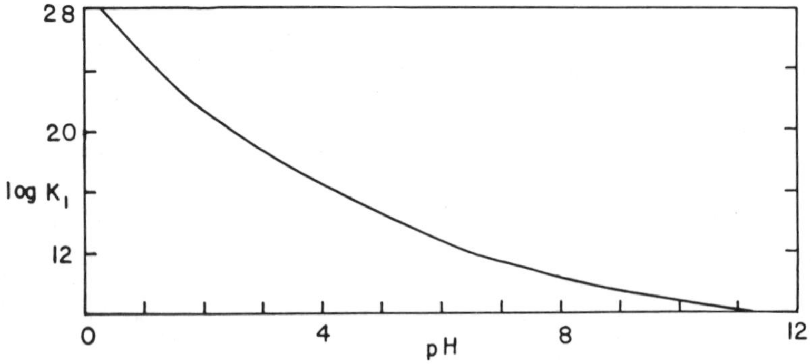

Figure 9-7. The minimum pH for quantitative titration of metal ions by EDTA at 0.01 M final C_M and 99.9% complexed.

Example 4. Show why EDTA titration is found to work for $Hg(NO_3)_2$ solutions under slightly acid conditions, but does not work if chloride is present. At what p**H** should EDTA complex Hg(II) if unbound chloride is 0.01 *M*?

We showed above that Hg^{2+} is complexed quantitatively by EDTA down to p**H** 1.9. However, chloride competes for the Hg^{2+}. We need to find an α_{0Hg} in the chloride system (Figure 9-2). Since $HgCl_2$ is the major species here, we can approximate the total [Hg(II)] in the Cl^- system as $[HgCl_2]$. Then we use β_2,

$$\beta_2 = 10^{13.22} = \frac{[HgCl_2]}{[Hg^{2+}][Cl^-]^2} \qquad \text{and} \qquad \alpha_0 \cong \frac{[Hg^{2+}]}{[HgCl_2]}$$

So, for our conditions, α_0 is $10^{-9.22}$. Putting these values into the K_f expression, we get

$$K_f = 10^{21.80} = \frac{[HgY^{2-}] - x}{x\alpha_{0,Hg}x\alpha_{0Y}} = \frac{0.01}{10^{-5}10^{-9.22}10^{-5}\alpha_{0Y}}$$

We took $x = 10^{-5}$ for the quantitative condition at equivalence as in previous examples. This yields $\alpha_{0Y} = 10^{-4.58}$, which, from Figure 9-5, occurs at p**H** 6.0. In practice, p**H** 10 with NH_3, and basic indicators, is used to overcome chloride interference in mercury(II) titrations.

Examination of the formation constants for the various complexes in Table 8-2 allows prediction of likely ways to mask one metal ion to permit titration of another by EDTA:

(a) The hard acid Al^{3+} can be masked by F^- to permit EDTA titration of Hg^{2+}, a soft acid, which is little removed by F^-.

(b) The same pair might be resolved by complexing the mercury ion by the soft base I^-, which does not complex Al^{3+}.

5. Amino Acid Complexes of Copper and Cobalt Ions

α-Amino acids are good ligands in basic solution for many metal ions. The amino acid anion is usually bidentate so that Cu^{2+} holds two, and cobalt(II) holds up to three, of these ligands. The bright-blue copper complex has been made the basis of spectrophotometric and titrimetric methods for amino acid determination. For glycine, for

example, we have

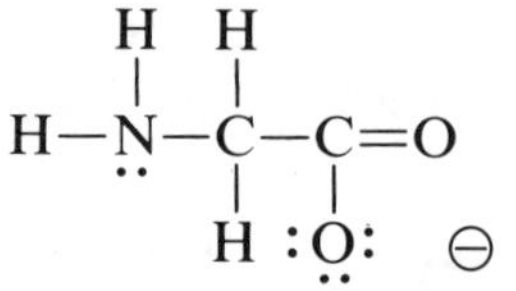

$pK^\circ_{a_1} = 2.35, \qquad pK^\circ_{a_2} = 9.78$

(shown is the *glycinate ion*, which forms a five-membered ring with M^{2+}). The formation constants for the copper ion complexes are $\log K_1 = 8.23$ and $\log K_2 = 6.96$ at $I = 0.1\ M$ and at 25°. By now the reader can visualize the α diagram much like Figure 8-2 which these constants will produce with crossings at log [glycinate] = -8.23 and -6.96. Conditions for quantitative complexing by Cu^{2+} should be easy to arrange.

With Co^{2+}, reported constants with glycinate are $\log K_1 = 4.65$, $\log K_2 = 3.78$, and $\log K_3 = 2.38$ at $I = 0.15$ and at 25.1°. The α diagram for cobalt–glycinate is shown as Figure 9-8.

Suppose the **pH** of such a system is varied. How is the complexing of glycinate affected? Let us examine the **pH** dependence, starting at **pH** 11 with total glycine 0.20 *M* and total cobalt 0.010 *M*. Figure 9-8

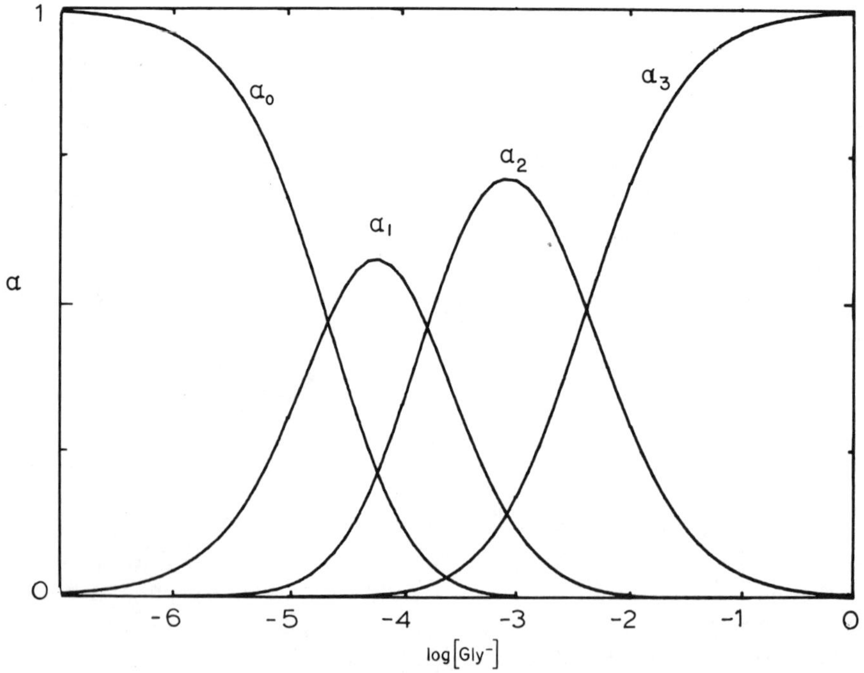

Figure 9-8. α curves for the system Co(II)–glycinate.

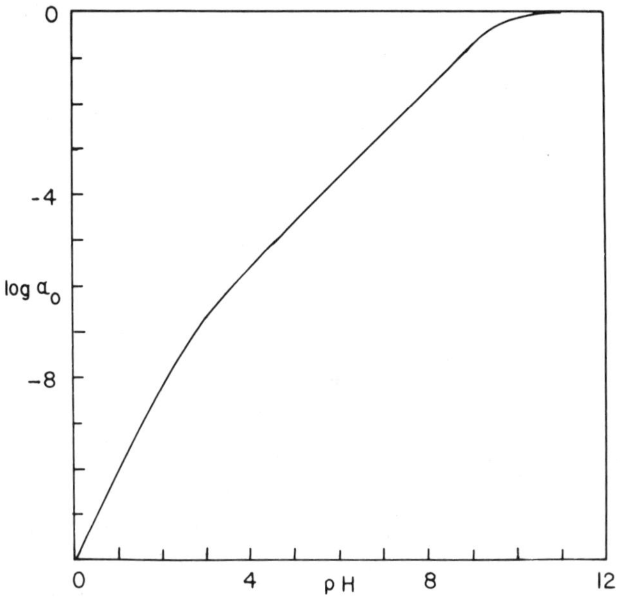

Figure 9-9. Log α_0 curve for glycine.

tells us that about 98% of the cobalt is as $Co(Gly)_3^-$ ion, and we know that glycine must be largely the anion at this **pH** (Figure 6-2). To one-figure approximations, we can take glycine unbound by Co(II) as 0.2 *M*. As acid is added, **pH** lowered, the glycinate ion concentration falls and the proportions of cobalt complexes shift to the left in Figure 9-8. We can use Figure 9-9 and calculate the glycinate ion concentration at a series of **pH** values since $[Gly^-]$ is $0.2\alpha_0$. We then can read the α_n values of the cobalt species on Figure 9-8. These are plotted in Figure 9-10. It is similar to Figure 9-8 because of the nearly linear relation of log $[Gly^-]$ to **pH**. Since we are limited to a maximum $[Gly^-]$ of 0.2 *M*, the high **pH** limits of the α_n values are governed by this. (In K_3 we get $\alpha_3/\alpha_2 = 48/1$ as the maximum α_3 possible under these conditions.)

6. Weak Complexing at Moderate Concentrations: The Sea Water Problem

As an important practical case of multiple, homogeneous equilibria, consider sea water. The analytical concentrations of major ions are given in Table 9-5. Ionic strength is about 0.7 *M*, pH 8.15

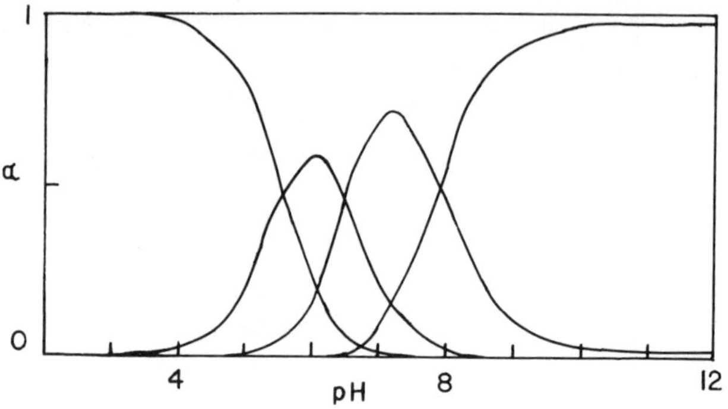

Figure 9-10. pH dependence of complexing in a solution of analytical concentrations, 0.20 *M* glycine and 0.010 *M* Co(II).

Table 9-5. Analytical Concentrations of Major Ions in Sea Water at 25°

Ion	*M*	Ion	*M*
Na^+	0.48	Cl^-	0.56
K^+	0.010	SO_4^{2-}	0.028
Mg^{2+}	0.054	HCO_3^-	0.0024
Ca^{2+}	0.010	CO_3^{2-}	0.00027

(at 25°), and temperature averages about 5°C. The question of what associations occur and what are the species concentrations has only recently been deduced from a serious attempt to consider all relevant equilibria. We follow the approach of Garrels and Christ.[3] For our example calculation, we treat the sulfates only.

Experimental determination of the formation constants of possible complexes among all the pairs (16) in Table 9-5 at ionic strength 0.7 is difficult. The set of constants used by Garrels and Christ is given in Table 9-6. Association by chloride seems negligible. Since the other anions have such low concentrations, not much of the cations can be associated even if most of the anions are tied up with them. This simplifies the calculation: For a fair approximation we may assume that the concentrations of the cations are as given in Table 9-5. The

[3]R. M. Garrels and C. L. Christ, *Solutions, Minerals, and Equilibria*, Harper and Row, New York, 1965.

Table 9-6. Log K_1° for Association between Ions[a]

Ion	HCO_3^- (0.68)	CO_3^{2-} (0.20)	SO_4^{2-} (0.12)
K^+ (0.64)	—	—	0.96
Na^+ (0.76)	−0.25	1.27	0.72
Ca^{2+} (0.28)	1.26	3.2	2.31
Mg^{2+} (0.36)	1.16	3.4	2.36

[a] Activity coefficient of each ion given in parentheses.

set of equations from the constants and a material balance serves to determine the species. Let us show details for the sulfates.

We can write four equilibrium conditions for sulfate and one material balance to obtain five equations in five unknowns. Let x be the equilibrium sulfate concentration (unbound), and a, b, c, and d the concentrations of the sulfate complexes of Na^+, K^+, Ca^{2+}, and Mg^{2+}. Using the K's and activity coefficients, we obtain

$$10^{0.72} = 5.2 = \frac{f_-[NaSO_4^-]}{f_+[Na^+]f_{2-}[SO_4^{2-}]} = \frac{0.68a}{0.76(0.48 - a)(0.12)x}$$

$$10^{0.96} = 9.1 = \frac{f_-[KSO_4^-]}{f_+[K^+]f_{2-}[SO_4^{2-}]} = \frac{0.68b}{0.64(0.01 - b)(0.12)x}$$

$$10^{2.31} = 204 = \frac{(CaSO_4)}{f_{2+}[Ca^{2+}]f_{2-}[SO_4^{2-}]} = \frac{c}{0.28(0.01 - c)(0.12)x}$$

$$10^{2.36} = 229 = \frac{(MgSO_4)}{f_{2+}[Mg^{2+}]f_{2-}[SO_4^{2-}]} = \frac{d}{0.36(0.054 - d)(0.12)x}$$

$$\text{balance:} \qquad [\text{total sulfate}] = 0.028 = a + b + c + d + x$$

Assuming the subtractive terms in the denominators to be small for a first approximation allows easy substitution of x for all the other unknowns to obtain $x = 0.0144$. Returning this to the other equations gives us,

$$a' = 0.0050, \qquad b' = 0.00015, \qquad c' = 0.0010, \qquad d' = 0.0077$$

We see that the first two are satisfactory for omission from subtraction in the first approximation made (beyond the two significant figures available) but the last two are significant. These can be inserted in the

Table 9-7

Ion	M (total)	% free	% Ca complex	% Mg complex	% Na complex	% K complex
SO_4^{2-}	0.028	54	3	22	21	0.5
HCO_3^-	0.0024	69	4	19	8	—
CO_3^{2-}	0.00027	9	7	67	17	—

denominator terms to obtain a second, more satisfactory approximation, $x = 0.0150$ and

$$[NaSO_4^-] = 0.0050, \qquad [KSO_4^-] = 0.00015$$

$$[CaSO_4] = 0.0009, \qquad [MgSO_4] = 0.0068$$

This means that 54% of the sulfate is left uncomplexed.

The conclusion of all the calculations by Garrels and Christ is given in Table 9-7. The percentages of cations not complexed are

$$Na^+ \ 99, \qquad K^+ \ 99, \qquad Mg^{2+} \ 87, \qquad Ca^{2+} \ 91$$

This new picture of sea water allows rational explanations of its **pH** and the solubilities of several compounds, including the important $CaCO_3$ and $CaSO_4$, which we shall discuss in detail in the next chapter using these results.

7. Calculation of Equilibrium Constants from Data

Examination of the relations associated with equations (8-5) and (8-6) shows that constants of the system should be calculable if one can determine the master variable, the unbound ligand concentration. This allows one to calculate, from the analytical concentrations, the value of $\bar{n}$, the average number of ligands bound per metal ion. In general, having N values of $\bar{n}$ in N different mixtures allows determination of the N formation constants. N is the maximum number of ligands accepted by the metal ion, and also the number of step formation constants.

For example, Bjerrum[4] reported the following results for the silver(I)–NH_3 system at 22° in 2 M NH_4NO_3 medium. Measurement

[4] J. Bjerrum, *Metal Ammine Formation in Aqueous Solution*, Haase and Son, Copenhagen, 1957.

of pH leads to a value of the unbound NH_3 and $\bar{n}$ for each mixture. Details of the laboratory procedure for what is called Bjerrum's method are given in his book and elsewhere[5]:

unbound $[NH_3]$	1.094×10^{-4}	4.198×10^{-4}
$\bar{n}$	0.492	1.463

From equation (8-6) we have

$$\bar{n} = (1 - \bar{n})K_1[NH_3] + (2 - \bar{n})K_1K_2[NH_3]^2 \tag{9-3}$$

$$K_2 = \frac{\bar{n} - (1 - \bar{n})K_1[NH_3]}{(2 - \bar{n})K_1[NH_3]^2}$$

Substituting the data into this and setting the two expressions for K_2 equal to each other allows calculation of K_1 and then K_2:

$$K_1 = 10^{3.362}, \qquad K_2 = 10^{3.943}$$

In practice, more than two mixtures are measured, and graphical or computer reduction of the data yields best values for the constants. For this case with $N = 2$, we can rearrange equation (9-3),

$$\frac{\bar{n}}{(1 - \bar{n})[NH_3]} = K_1 + \frac{(2 - \bar{n})[NH_3]K_1K_2}{(1 - \bar{n})}$$

Now, a plot of the left side vs. the factor $(2 - \bar{n})[NH_3]/(1 - \bar{n})$ for a series of mixtures should produce a straight line of slope K_1K_2 and intercept K_1. An example of this approach is found in the report on the Cu(II)–F^- system by Gifford.[6]

When the system has N of 3 or more, a similar treatment will give valid results only if the K values are in descending order and separated by a power of ten or more. This means that three species and two adjacent K's describe the system well at any given log [L] value. See Figures 8-3–8-5. In the less favorable cases, such as Cu(II)–NH_3 (Figure 9-1), groups of three K's can be used in the $N = 3$ form from equation (8-6). The experimental $\bar{n}$ curve is obtained just as with Ag(I)–NH_3. First approximations to the K values can be read

[5] W. B. Guenther, "Stepwise Formation Constants," *J. Chem. Ed.* **44**, 46 (1967); *Quantitative Chemistry*, Addison-Wesley, Reading, Massachusetts, 1968, "Experiments," pp. 355–361.
[6] S. Gifford, *Inorg. Chem.* **13**, 1434 (1974).

from this curve at the half-integral $\bar{n}$ values. This is true because of the effects of various values of α on $\bar{n}$, as follows. Refer to Figure 9-1. The value $\bar{n} = 0.5$ can occur when $\alpha_0 = \alpha_1 \cong 0.5$. If this is true, $[Cu^{2+}] = [CuNH_3^{2+}]$, and in the K_1 expression this gives us

$$K_1 = \frac{[CuNH_3^{2+}]}{[Cu^{2+}][NH_3]} = [NH_3]^{-1}$$

At $\bar{n} = 0.5$ we read $[NH_3] = 10^{-4.40}$ on Figure 9-1.

Next, the values $\alpha_1 = \alpha_2 \cong 0.5$ give $\bar{n} = 1.5$. This works even if adjacent α values are not negligible, provided $\alpha_0 \cong \alpha_3$. This occurs at $K_2 = [NH_3]^{-1} = 10^{3.58}$. Similarly, we read $K_3 = 10^{2.92}$ at $\bar{n} = 2.5$; $K_4 = 10^{2.05}$ at $\bar{n} = 3.5$; $K_5 = 10^{-0.5}$ at $\bar{n} = 4.5$; and $K_6 = 10^{-2}$ at $\bar{n} = 5.5$. Clearly these are not the set of K's used to produce these curves and the α curves show us why (Figure 9-1). At $\bar{n} = 0.5$, α_2 is not negligible. But, at $\bar{n} = 1.5$, $\alpha_1 = \alpha_2$ and $\alpha_0 = \alpha_3$ while α_4, α_5, and α_6 are practically zero. This does give us $\bar{n} = 1.50$ and a good K_2 value. This happens again for K_3. But, at $\bar{n} = 3.5$ we see that $\alpha_2 > \alpha_5$, and so on for the rest of the K's which are not reliable. Nevertheless, these approximate K values serve to start the refining process. At $\bar{n} = 0.5$, we can use equation (9-3), putting in the rough K_2 and solving for a better K_1. Then the three-K ($N = 3$) equation can be used to obtain better values of K_1 and to check K_2 and K_3. Then we proceed to successive groups of three K's. The calculations are long, so that rapid computers are useful. Varga[7] describes a program for this purpose and illustrates it with the earlier data of Bjerrum on the Cu(II)–NH_3 system.

The making of the mixtures for this method of K determination is often achieved by a titration-like process. The ligand solution may be added, or a strong base may be added to the solution of metal ion, protonated ligand, and some strong acid. Thus, the equilibria

$$HL \rightleftharpoons H^+_{aq} + L, \qquad M + L \rightleftharpoons ML, \qquad \text{etc.}$$

are shifted at each addition of the base. A pH curve obtained without the metal ion present serves to establish the equilibrium quotient of HL under the conditions used.

Solvent extraction as a competing equilibrium provides a means of determining the equilibrium concentration of one or more species

[7] L. P. Varga, *Anal. Chem.* **41**, 323 (1969).

in a mixture. The mercury(II) halide systems have been reinvestigated via benzene extraction, by Marcus.[8] The benzene extracts only the neutral molecules like $HgCl_2$, so that once the constant for extraction,

$$P = \frac{[HgCl_2]_{(bz)}}{[HgCl_2]_{(aq)}}$$

has been determined, one can calculate the $[HgCl_2]$ in various mixtures by means of a mercury determination of the benzene extract.

8. Some Physiological Reactions of Metal Ions

The metal chelates of EDTA are quite inert in the body and are easily excreted through the kidneys. This gives EDTA valuable uses in medicine.

For acute lead poisoning, the treatment is intravenous administration of 3% $Na_2CaEDTA$ in isotonic saline or glucose solution. The equilibrium constant for the metal ion interchange is

$$CaY^{2-} + Pb(II) \rightleftharpoons PbY^{2-} + Ca^{2+}, \qquad K_{eq} = K_{Pb}/K_{Ca} = 10^8$$

using the K values from Table 9-2, rounded off. Thus, the lead complex should form and be excreted as found. EDTA alone would remove excessive Ca^{2+} from the body, so the calcium EDTA form is used here. Rapid relief from the poisoning symptoms takes place. However, after several days or weeks they may return because lead deposited as phosphates, etc., in bone is slowly released. The EDTA treatment is repeated until final recovery is achieved. Similar EDTA treatment has been effective in removing a variety of radioactive metal ions from the body. This is expected due to the high formation constants in Table 9-2 for most heavy metal ions.

MgEDTA injected into the bloodstream increases Mg^{2+} at the expense of Ca^{2+} and produces interesting effects for study; slower heart rate and lower blood pressure. Here the relevant K_{eq} is just favorable, as above, $K_{eq} = 10^2$.

Iron(III)EDTA is used extensively in correcting iron deficiency in plants. It is soluble where $Fe(OH)_3$ would precipitate from ionic

[8] Y. Marcus, *Acta Chem. Scand.* **11**, 329, 599 (1957).

Fe(III) solution.

$$Fe(OH)_3 + Y^{4-} \rightleftharpoons FeY^- + 3OH^-$$

$$K_{eq} = \frac{[FeY^-][OH^-]^3}{[Y^{4-}]} = K_{so}K_{1(Fe)} = 10^{-37}10^{25} = 10^{-12}$$

Thus at **pH** 7, $[OH^-] = 10^{-7}$, we find the ratio

$$[FeY^-]/[Y^{4-}] = 10^{-12}/(10^{-7})^3 = 10^9$$

Thus, the Fe(III) should remain in solution. The plants seem able to get the Fe(III) from this soluble complex.

The poisoning by heavy metal ions seems to result from their complexing of Lewis base positions on nucleic acids, enzymes, etc., which are then prevented from their normal operation. EDTA treatment is useful for the cases mentioned above. However, mercury ions in the body must be present as $HgCl_2$, and other complexes which are not favorable for EDTA reaction. For example, take the case of $HgCl_3^-$,

$$HgCl_3^- + CaY^{2-} \rightleftharpoons HgY^{2-} + Ca^{2+} + 3Cl^-$$

$$K_{eq} = K_{1(HgY)}/K_{1(CaY)}\beta_{3(HgCl_3)} = \frac{[HgY^{2-}][Cl^-]^3[Ca^{2+}]}{[HgCl_3^-][CaY^{2-}]} = 10^{-3}$$

Similar unfavorable constants can be obtained for the other cases. A successful agent in complexing the Hg(II) is BAL (British anti-Lewisite), 2,3-dimercaptopropanol, with potent sulfur ligands for the mercury, and also for arsenic, antimony, bismuth, and gold. It is injected intramuscularly in 10% solution in peanut oil at a dose rate of 3 mg/kg of body weight every 4 hr for two days, and after that at decreasing frequency until recovery.

Possible Equilibria in Blood Fluids. Major constituents in human blood, which is at pH 7.36 and a temperature of 37°C, are:

Na^+ (*S*)	0.140 *M*	Amino acids (*P*)	0.003 *M*
Cl^- (*S*)	0.104 *M*	Glucose	0.005 *M*
HCO_3^-	0.028 *M*	Lactic acid	0.0013 *M*
Ca^{2+} (total) (*S*)	0.0025 *M*	Pyruvic acid	1×10^{-4} *M*
Ca^{2+} (ionic) (*S*)	0.001 *M*	Protein, total (*P*)	7.2 g/100 ml
		Hemoglobin	15 g/100 ml
		Fats, total (*P*)	0.74 g/100 ml

These are for whole blood unless noted *S*, for serum (the centrifugate of clotted blood, most protein removed), or *P*, for plasma (the centri-

Table 9-8

Blood component	Amino acid	Cl^-	Lactate	Pyruvate	HCO_3^-
Concentration, M	0.003	0.104	0.0013	10^{-4}	0.028
log K_1 toward Hg(II)	10.3 (glycine)	6.74	5 (?)	5 (?)	?

fugate of blood after anticoagulant treatment with heparin or oxalate). The HCO_3^- is for venous blood and is in equilibrium with small amounts of $H_2CO_3(CO_{2aq})$ and CO_3^{2-}.

Of interest here is the question, What form should be assumed for the metal ions introduced into the circulatory system? To answer this, one must consider all the possible ligands present. For mercury(II), one may summarize these with their formation constants as shown in Table 9-8. In addition, the large amount of protein surface present will offer ligand sites like the glycine, or like cysteinate (with an –SH group), which has a K_1 toward Hg^{2+} of 10^{14}. So we may confidently predict that mercury ion will be largely bound to Cl^- and the organic sites in blood. Thus, a much better ligand than EDTA will be required to remove it from the system. The 2,3-dimercaptopropanol used probably has a K_1 for Hg^{2+} even above that for the cysteine case.

We have considered only some of the possibilities for complexing. The phosphate groups present in many molecules also bind metal ions. Much of the Ca^{2+} and Mg^{2+} of normal blood is so bound up. The lack of formation constants prevents a quantitative treatment of any one metal ion. However, one can make some approximations, as for the Hg^{2+} case, which are far better than assuming that the total analytical concentration of the metal in blood is present as its aquo ion.

The widespread importance of trace metal ion complexing in biochemical systems has become apparent from recent research. Iron and magnesium in hemoglobin and chlorophyll are well known, while the need and functions of Co, Zn, Mn, Cr, Mo, and others are under active investigation. Knowledge of possible competing equilibria is required for understanding of reactions of these metal ions.

Problems

1. Examine the α diagrams and then estimate, without calculations, what major species of the metal ion are present in mixtures having the following analytical concentrations of M and L.

a. 0.10 M Mg^{2+} and 1.00 M NH_3.
b. 0.10 M Co^{2+} and 1.00 M NH_3.
c. 0.100 M Cu^{2+} and 0.200 M NH_3.
d. 0.010 M Hg^{2+} and 0.020 M NH_3.
e. 0.010 M Hg^{2+} and 0.010 M NH_3.
f. 0.100 M Hg^{2+} and 1.00 M Cl^-.
g. 0.100 M Hg^{2+} and 0.010 M Cl^-.

2. Use the approximate results above to refine the numerical α values either with K expressions or by the $\bar{n}'$ method.

3. Calculate the ratio of complexed to aquo Pb^{2+} after a poisoning victim is injected with enough $CaEDTA^{2-}$ to make its concentration 10^{-5} M in the serum, which has about 10^{-3} M Ca^{2+}.

4. Given the data in Problem 3, and that serum $[Cl^-]$ is 0.10 M, calculate the ratio $[HgY^{2-}]/[HgCl_3^-]$ after ingestion of Hg(II).

5. Given the serum levels of Cl^- and glycine as 0.10 M and 0.003 M, and α_0 for glycine as about 0.01, calculate the ratio of $[HgGly^+]/[HgCl^+]$ when trace Hg(II) is present (no EDTA).

6. At what **pH** should 10% of the lanthanum–EDTA complex be dissociated in 0.020 M equivalent $LaEDTA^-$ solution?

7. What should be complexed and what volume of 0.0200 M EDTA will be required for 25.0 ml of sea water titrated at (a) **pH** 11, (b) **pH** 7?

8. How many moles of HCl gas must be added to one liter of 0.100 M $Hg(NO_3)_2$ to make the average complex $HgCl_3^-$, $\bar{n} = 3.00$?

10 | Solubility Equilibria

The equilibrium between a solid and its saturated aqueous solution is rarely as simple as the acid–base and M–L homogeneous cases we have treated. We begin with the simplest examples and then introduce the various types of competing equilibria one by one.

1. Simple Solubility: No Competing Reactions

The solubility of a molecular material represents the simplest kind of equilibrium, having only one species on each side of the equation:

$$A_{(c)} \rightleftharpoons A_{(aq)} \qquad \text{and} \qquad K^\circ = (A)/(A)_{(c)}$$

Since the pure solid $A_{(c)}$ has a constant activity, it follows that the activity of the dissolved material in equilibrium with it is also constant, at a given temperature. In general, when ionic strength and solubility are both not much above 0.1 M, activities of uncharged species can be taken equal to their molarities for accuracy within 2%. The solubility of such compounds as benzene, ethyl acetate, etc., in water is then the equilibrium constant for this reaction.

Only a few ionic compounds have both low solubility and ions of low acid–base and complexing reactivity. When such complications are absent, the only equilibrium is

$$M_aX_{b(c)} \rightleftharpoons aM + bX \qquad \text{and} \qquad K^\circ = \frac{(M)^a(X)^b}{(M_aX_b)}$$

The pure solid has constant activity, so that chemists have come to use the ion solubility product to express the relationship of the variables

$$K^\circ_{s0} = (M)^a(X)^b = f^a_+[M]^a f^b_-[X]^b \qquad (10\text{-}1)$$

A temperature of 25°C is understood unless another is stated. Solubility and K_{s0} often increase rapidly with temperature. The significance of K_{s0} is to relate the solid equilibrium with the ions (zero complex), K_{s_1} is for a first complex, solid to MX, etc. The K_{s0} is then the traditional K_{sp}. The K_{s_1}, K_{s_2}, etc., will be illustrated later in the chapter.

What kinds of substances should satisfy this simple equation? Strong electrolytes with ions of low acid–base strength are likely candidates. Many electrolytes which were once treated by this simple equation have been found to be weak electrolytes, and many form even higher complexes than the first. For example, AgCl dissolves in HCl solutions as AgCl molecules and complexes $AgCl_2^-$, $AgCl_3^{2-}$, etc. as well as in simple ionic form. Most ions of charge greater than one also associate with oppositely charged ions in solution. We shall deal with these later. Thus, it seems that we are limited to singly charged ions which are large and unreactive, namely, the heavier alkali metal ions, perchlorate, and a few others.

There is a further requirement if we want to be able to apply Debye–Hückel theory to correction for activity effects. That is the necessity for an unchanging medium, for some purposes, nearly pure water. Thus, the solubilities must be small in order to keep dilute solutions. For example, pure water is 55.4 *M* in H_2O; 0.1 *M* NaCl is about 55.2 *M* in water; but saturated NaCl (5.4 *M*) is only about 49 *M* in H_2O. In 0.1 *M* NaCl we have about 280 H_2O molecules for each ion, Na^+ or Cl^-, but in the saturated solution there are but four H_2O for each ion. In the first, the ions are well screened from each other, while in the saturated solution they are not. Fortunately for theoretical purposes there are several compounds made up of nearly ideal ions and which have low solubility. Most useful among these is $RbClO_4$ with water solubility 0.0683 *M* at 25°.

The solubility of $RbClO_4$ in the presence of some other strong electrolytes has been found to agree well with that expected from equation (10-1) when Debye–Hückel theory is used to obtain activity coefficients.[1]

First, we give results for the common ion effect at a constant ionic strength. For this condition, it is convenient to rearrange equation

[1]W. B. Guenther, *J. Am. Chem. Soc.* **91**, 7619 (1969).

Table 10-1. Solubility of $RbClO_4$ at Constant $I = 0.1\ M$[a]

Medium	$RbClO_4$ dissolved	$[Rb^+]$	$[ClO_4^-]$	K_{so}
0.0300 NaCl	0.0708	0.0708	0.0708	5.01×10^{-3}
0.0492 RbCl	0.0505	0.0997	0.0505	5.03×10^{-3}
0.0500 $HClO_4$	0.0500	0.0500	0.1000	5.00×10^{-3}

[a] At 25°. All values are *M*.

(10-1) to read

$$K_{so}^\circ/f_+f_- = K_{so} = [Rb^+][ClO_4^-] \qquad (10\text{-}2)$$

Three solutions were saturated with $RbClO_4$ to test this equation (Table 10-1). The NaCl has no ion in common with $RbClO_4$, but only serves to adjust the ionic strength to 0.1 *M*. Notice how an increase in one of the reacting ions, $[Rb^+]$ or $[ClO_4^-]$, results in a decrease in the other just sufficient to keep the K_{so} constant within experimental reliability.

Second, we look at data with varied ionic strength. In Table 10-2 the ionic strength is used to calculate f values from the Debye–Hückel equation (3-2), and from them, K_{so}° from equation (10-2). Notice that K_{so} increases with I, but that the calculated K_{so}° is constant within experimental reliability.

It is often useful to examine data graphically. The choice of just what functions of the data are plotted will determine how useful the graphs will be. By looking at equation (10-2), rearranged into slope–intercept form for a straight line, one may decide to plot the $[Rb^+]$ vs. $[ClO_4^-]^{-1}$ because the slope of this straight line will give K_{so} and the intercept should be 0,0. If the line is not straight, or if there is a positive intercept, the data are telling us that the simple equilibrium we have assumed is not the only one operating (assuming accurate data), which is useful information. To clarify the linear relationship, look at the slope–intercept (*m–b*) formulation of a straight line (for the constant *I* cases),

$$y = mx \qquad\qquad + b$$

$$[Rb^+] = K_{so}[ClO_4^-]^{-1} + 0$$

The data follow this relation as closely as the experimental precision can reveal (Figure 10-1). We shall soon meet systems which deviate widely from this behavior.

Table 10-2. Solubility of $RbClO_4$ at Varied Ionic Strengths[a]

Medium	$RbClO_4$ dissolved	$[Rb^+]$	$[ClO_4^-]$	$K_{so} \times 10^3$	I	$\frac{\sqrt{I}}{1+\sqrt{I}}$	$f_{\pm}$	$K_{so}^{\circ} \times 10^3$
Pure H_2O	0.0683	0.0683	0.0683	4.66	0.0683	0.207	0.784	2.88
0.0200 $LiClO_4$	0.0602	0.0602	0.0802	4.83	0.0802	0.220	0.772	2.89
0.0492 RbCl	0.0505	0.0997	0.0505	5.03	0.0997	0.240	0.775	2.87
0.0504 NaCl	0.0726	0.0726	0.0726	5.26	0.1230	0.259	0.737	2.88
0.1000 $HClO_4$	0.0386	0.0386	0.1386	5.35	0.1386	0.271	0.727	2.85

[a] At 25°. All values are *M*.

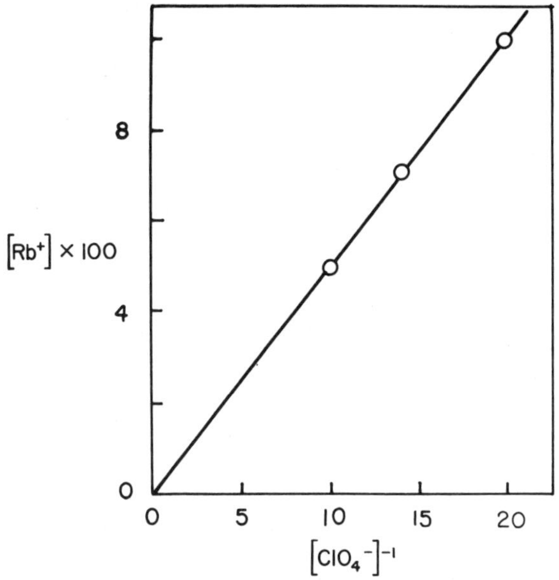

Figure 10-1. The common-ion effect on the solubility of $RbClO_4$ at constant ionic strength, 0.1 *M*, at 25°.

To test the application of Debye–Hückel theory, take the logs of equation (10-2) and arrange for linear plotting,

$$y = mx \qquad\qquad + b$$

$$\log K_{s0} = 2(0.51)\sqrt{I}/(1 + \sqrt{I}) + \log K_{s0}^{\circ}$$

Thus, a plot of $\log K_{s0}$ vs. $\sqrt{I}/(1 + \sqrt{I})$ should produce a straight line of slope 1.02 and intercept $\log K_{s0}^{\circ}$ at $I = 0$. For the f values in equation (10-2), we have used equation (3-2). Figure 10-2 has intercept -2.542, or $K_{s0}^{\circ} = 2.87 \times 10^{-3}$, in good agreement with Table 10-2.

In so long an extrapolation as done here, the uncertainty of the result may be large. Here, the error is reduced by assuming the known Debye–Hückel slope.

Once satisfied that equation (10-2) correctly represents the solubility equilibrium, we can calculate solubilities in a variety of solutions when the appropriate K_{s0} values are available.

Example 1. Calculate the solubility of $RbClO_4$ in 0.0400 *M* $HClO_4$. Let the solubility be x moles per liter. Assuming constant volume, we have, after equilibrium is reached, $[Rb^+] = x$, $[ClO_4^-] =$

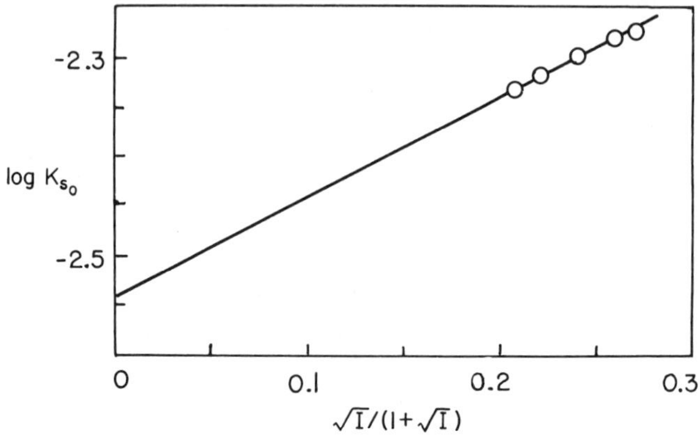

Figure 10-2. $RbClO_4$ data following a Debye–Hückel slope.

$0.0400 + x$, and $I = 0.0400 + x$. Since the ionic strength must be well above 0.04 *M*, we can see that the effective K_{s0} must be significantly larger than K_{s0}°. Let us try, for a first approximation, 4×10^{-3}. [We must approach by this successive approximation method because there are no direct ways to solve the kind of equation one gets by substituting the Debye–Hückel expression for $\log f_{+}$ into equation (10-2). The student may wish to see this by using $I = 0.0400 + x$. The resulting equation must be solved by successive approximations.] This problem comes up here only because the solubility is so great that the final ionic strength is quite different from 0.04 *M*. We have

$$K_{s0} = [Rb^{+}][ClO_4^{-}]$$

$$4 \times 10^{-3} = x(0.0400 + x) \qquad \text{or} \qquad x^2 + 0.04x - 0.004 = 0$$

The quadratic formula gives

$$x = -\frac{0.04 \pm (0.0016 + 0.016)^{1/2}}{2} = 0.046\ M$$

and

$$I = 0.086\ M$$

(Only the positive root has physical significance in these problems.) One then calculates a better effective K_{s0} value for this ionic strength with the Kielland table (Appendix A-1) and continues until an unchanging solubility about 0.053 *M* is obtained.

2. Solubility with Complex Formation

Let us examine the experimental results on solubility of $CaSO_4$, which do not follow the simple relation that worked so well with $RbClO_4$. In 1929, some careful determinations were made of the solubility of $CaSO_4$ in a series of dilute solutions of common ions, some with $CaCl_2$ and others with K_2SO_4 present. Until recently it was impossible to interpret these results. First, consider the results with $CaCl_2$ given in Table 10-3. As a first, naive treatment, we assume roughly constant ionic strength and plot S vs. $[Ca^{2+}]^{-1}$ [equation (10-2)] to get line *a* of Figure 10-3. Since its intercept is rather far from 0,0, we decide that we should take ionic strength into account with the doubly charged ions. We get I from equation (2-4) (here it is $4[CaSO_4] + 3[CaCl_2]$), look up f values (tables in Appendix A-1), and calculate $([Ca^{2+}]f_{\pm}^2)^{-1}$ to plot, as in the previous example, line *b* in Figure 10-3. This still does not go to 0,0. A skeptic might argue that this is merely because of deviation from Debye–Hückel behavior at the higher ionic strengths, and point out that the dashed line through the more dilute points does give a line to (0,0). This reason is on the borderline of plausibility since the ionic strength of the first five data is 0.1 or less. The last four range from 0.12 to 0.2. Here the matter would rest until new data or a new idea about ion associations arrive to settle the question.

We look for omissions: Proton association with SO_4^{2-} can be ruled out using the K_a value, as can $CaOH^+$ formation. How would association of the doubly charged ions Ca^{2+} and SO_4^{2-} affect the situation? The solubility will be the material balance on sulfate,

$$S = [CaSO_4] + [SO_4^{2-}] \tag{10-3}$$

Table 10-3. Observed Solubility of $CaSO_4$ in $CaCl_2$ Solutions[a,b]

$CaCl_2$	S, $CaSO_4$	$CaCl_2$	S, $CaSO_4$	$CaCl_2$	S, $CaSO_4$
0	0.0153	0.0150	0.0118	0.0300	0.0104
0.0050	0.0137	0.0200	0.0112	0.0400	0.0097
0.0100	0.0126	0.0250	0.0108	0.0500	0.0093

[a] A. Seidell and W. F. Linke, eds., *Solubilities*, 4th ed., Van Nostrand, New York, 1958, Vol. I, p. 669.
[b] At 25°C. Values are *M*.

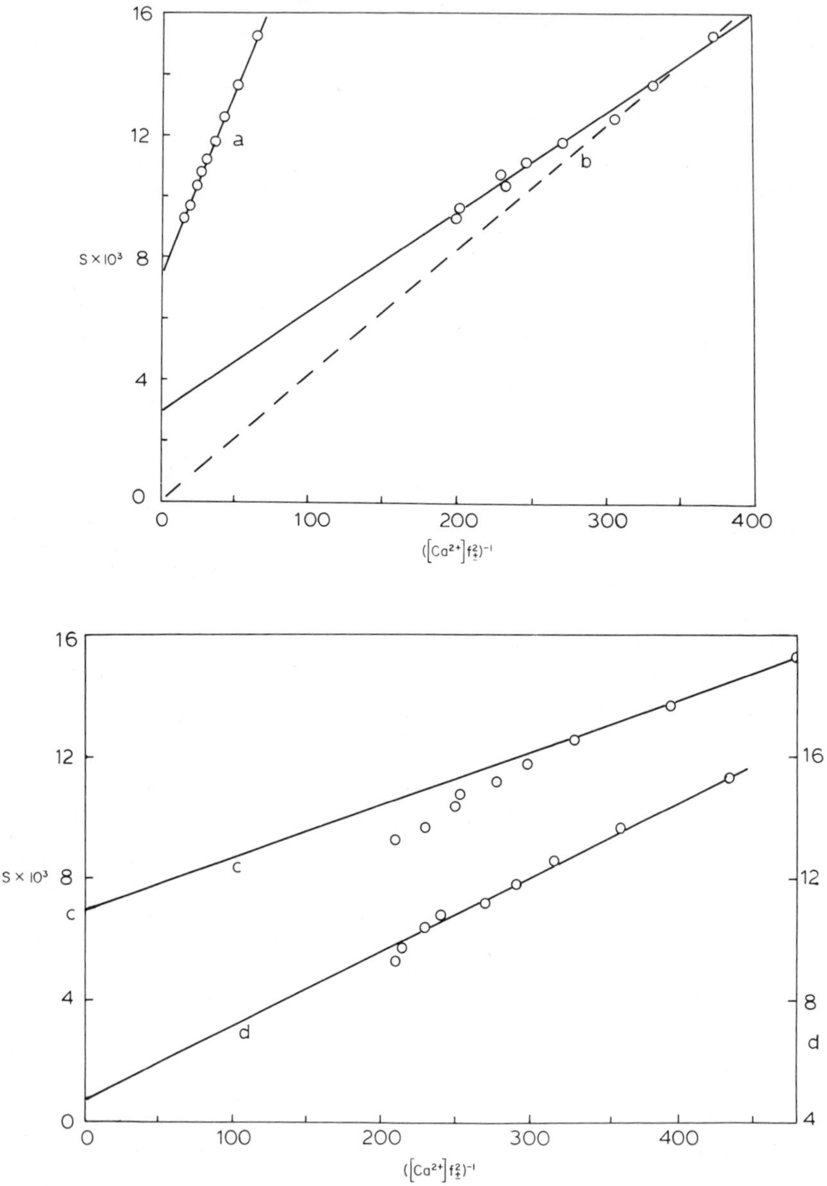

Figure 10-3. Plots of the solubility data for $CaSO_4$ in $CaCl_2$ solutions using (a) activity coefficients of 1; (b) activity coefficient values for strong electrolytes, (c) and subtracting molecular $CaSO_4$ as 0.007 *M*, (d) and subtracting molecular $CaSO_4$ as 0.0047 *M*. Note that the scale for (d) is at the right, displaced for clarity.

The two equilibria we now have are

$$CaSO_{4(c)} \rightleftharpoons CaSO_4 \qquad K^\circ_{s_1} = (CaSO_4)$$

$$CaSO_4 \rightleftharpoons Ca^{2+} + SO_4^{2-} \qquad K^\circ_{\text{diss}} = \frac{(Ca^{2+})(SO_4^{2-})}{(CaSO_4)}$$

The first is a case of simple molecular solubility as discussed at the start of the topic. The product of these two constants must still be a constant:

$$K^\circ_{s_1} K^\circ_{\text{diss}} = (Ca^{2+})(SO_4^{2-}) = K^\circ_{s0}$$

That is, the K°_{s0} is the equilibrium constant for the process of solid dissolving to form uncomplexed ions (zero complex), and $K^\circ_{s_1}$ refers to the solid dissolving to form the first complex. The K°_{s0} has been traditionally called *the solubility product*. It remains a valid constant in all ionic solutions under valid conditions, but it does not give any information about the other possible equilibria, nor a method for obtaining the solubility of the solid.

Substitution of these expressions converts equation (10-3) to

$$S = K^\circ_{s_1} + K^\circ_{s0}/([Ca^{2+}]f_\pm^2)$$

Thus, at last, we see why such plots as we made should have a nonzero intercept $K^\circ_{s_1}$, the constant concentration (activity) of $CaSO_4$ units in all *saturated* solutions of $CaSO_4$. However, now we must correct the data for the amount of Ca^{2+} and SO_4^{2-} not in ionic form. This gives new values of I and $f_\pm$. This gives a new plot, first using the intercept of plot *a*, 0.007 *M*, to make the corrections. This plot (*c*) is tested by drawing the line from the pure water point to the assumed intercept at 0.007,0. Again the three points of lowest ionic strength fall on the line. But we are not satisfied with the results otherwise. We quickly find out that assuming other trial intercepts is not a useful procedure; I is too insensitive to small changes in $K^\circ_{s_1}$. We turn to the literature for some independent method of determination of the degree of ion pairing of $CaSO_4$. Electrical conductivity has suggested values of about 0.005 *M* for the $CaSO_4$ concentration in the saturated solutions and some more recent work in 1967, which used a new kind of electrode that responds to the activity of Ca^{2+}, just as the glass electrode responds to a_{H^+}, gave a value of 0.0047 *M*.[2] Again

[2] F. S. Nakajama and B. A. Rasnick, *Anal. Chem.* **39**, 1022 (1967).

Table 10-4

KSCN, *M*	0.00548	0.01033	0.04133	0.04440	0.06662
$S \times 10^6$	1.62	3.65	30.0	33.6	79.9
KSCN, *M*	0.08885	0.1111	0.1334	0.1779	0.2224
$S \times 10^6$	139	238	356	724	1280

correcting the data with this value gives a final plot (*d*) which seems to accommodate the data well. This gives a slope of 2.46×10^{-5}, the K_{so}°.

Solutions with K_2SO_4 added agree well (see p. 186) with the result of our $CaCl_2$ calculations if corrections are made not only for $CaSO_4$ units but also for some KSO_4^- formation, $K_f \cong 10$. The results give a K_{so}° of 2.45×10^{-5}. So we see that some older but accurate data do agree with other measurements and the usually quoted K_{so}° value of 2.5×10^{-5} for the "solubility product" of $CaSO_4$.[3,4]

In general, then, one may conclude that a nonzero intercept may indicate weak electrolyte behavior in plots of data of this nature. The phenomenon is quite common. Furthermore, solids and ions which exhibit complexing to the first complex often go beyond to form higher complexes. We shall see a pronounced case of this in the silver thiocyanate solubility study considered next.

3. Multiple Strong Complexing

The silver halides, thiocyanate, and cyanide form a series of complexes with excess anion which can greatly increase the solubility over that in water alone. These compounds are of chemical interest because of their many uses in analysis. Figure 10–4 shows the great increase in solubility observed as KSCN is added to solutions saturated with AgSCN. Let us examine the data of Cave and Hume[5] to deduce what complexes form and to what extent. The pure-water solubility is indicated in Figure 10-4, as is the dashed line expected for simple K_{so} behavior without complexing. A constant high ionic strength was maintained with 2.2 *m* KNO_3 and the total dissolved silver(I), *S*, was determined by colorimetric, polarographic, or gravi-

[3] A. W. Gardner and E. Glueckauf, *Trans. Faraday Soc.* **66**, 1081 (1960).
[4] L. Meites, J. Pode, and H. Thomas, *J. Chem. Educ.* **43**, 667 (1966).
[5] G. C. B. Cave and D. N. Hume, *J. Am. Chem. Soc.* **75**, 2893 (1953).

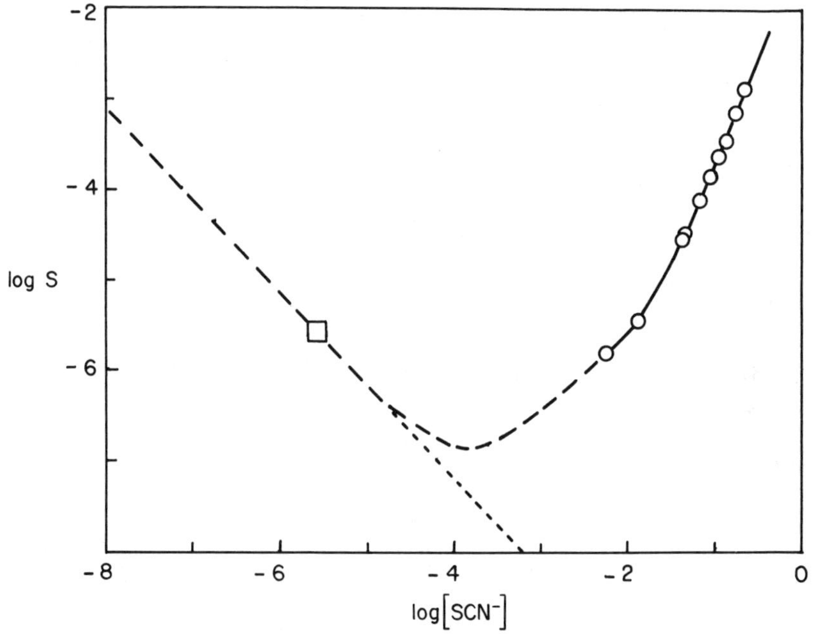

Figure 10-4. The solubility of AgSCN in KSCN solutions. Circles, data. Dashed line, calculated from equation (10-4). Straight extension for simple K_{so} behavior, no complexing. Square, pure water solubility.

metric means depending on the concentration. Some of the data are given in Table 10-4. The curve of several changes of slope suggests several stages of complexing. Hypotheses can be tested graphically following these authors' method. They found it necessary to go as far as four SCN^- per Ag^+. This leads to the solubility, material balance on total silver in solution [all possible forms of Ag(I)],

$$S = [Ag^+] + [AgSCN] + [Ag(SCN)_2^-] + [Ag(SCN)_3^{2-}] + [Ag(SCN)_4^{3-}]$$

It is convenient to express the equilibria involved by the expressions

$$K_{s_0} = [Ag^+][SCN^-], \qquad K_{s_1} = [AgSCN]$$

$$\beta_2 = \frac{[Ag(SCN)_2^-]}{[Ag^+][SCN^-]^2}, \qquad \beta_3 = \frac{[Ag(SCN)_3^{2-}]}{[Ag^+][SCN^-]^3},$$

$$\beta_4 = \frac{[Ag(SCN)_4^{3-}]}{[Ag^+][SCN^-]^4}$$

HSCN is a strong acid, so no K_a is needed. Let us abbreviate [SCN$^-$] as X, and use the K_{so} expression for [Ag$^+$] = K_{so}/X to get

$$S = K_{so}/X + K_{s_1} + \beta_2 K_{so} X + \beta_3 K_{so} X^2 + \beta_4 K_{so} X^3 \quad (10\text{-}4)$$

This is a solubility equation in terms of X alone. The ten pieces of solubility data above should be plenty to determine the five unknown constants in this equation. An effort to do this algebraically is instructive. K_{so} and K_{s_1} are so small that the precision of the data under the conditions makes it impossible to determine them well. Other data and extrapolation suggest that K_{so} is about 10^{-12} and that K_{s_1} is about 10^{-7} M. These mean that [Ag$^+$] and [AgSCN] are small for the data given, that only the last three terms in the S equation are important:

$$S \cong AX + BX^2 + CX^3$$

where A is $\beta_2 K_{so}$, B is $\beta_3 K_{so}$, and C is $\beta_4 K_{so}$. These constants were determined graphically. First rearrange to

$$S/X = A + BX + CX^2$$

A plot of S/X vs. X should be the positive half of a parabola going

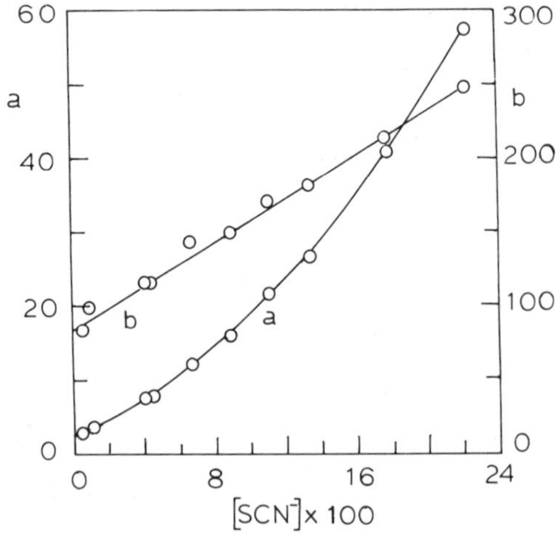

Figure 10-5. Calculational plots for AgSCN solubility study. Coordinate a is $10^4 S$/[SCN$^-$]. Coordinate b is (a − 2.5)/[SCN$^-$].

through $X = 0$ at $S/X = A$. This is shown as curve *a* in Figure 10-5. We get $A = 2.5 \times 10^{-4}$. Next rearrange again to

$$\frac{(S/X) - A}{X} = B + CX$$

A plot of the left side vs. X should be a straight line of intercept B and slope C. Curve *b* in Figure 10-5 shows this, giving $B = 0.0083$ and $C = 0.075$.

Their best value for K_{s0} at their conditions was 6.75×10^{-12}. This is combined with A, B, and C to get the β values for Ag^+–SCN^- complexing:

$$\beta_2 = 3.7 \times 10^7, \qquad \beta_3 = 1.2 \times 10^9, \qquad \beta_4 = 1.1 \times 10^{10}$$

Returning these to the complete solubility equation (10-4) gives the function to reproduce the data curve of Figure 10-4. We can calculate where the minimum should occur by differentiation. This gives $S_{min} = 1.4 \times 10^{-7}\ M$ at $[SCN^-] = 1.6 \times 10^{-4}\ M$.

We can now construct the log C diagram of the species (Figure 10-6). It turns out to consist entirely of straight lines, as do all similar cases (having saturation of the solid). The reason this must be so becomes apparent upon inspection of our log ratio methods (Chapters 5 and 8). In the present case, one soluble species, AgSCN, is constant, K_{s1}, and thus the ratios of the others to this have the linear log C relation. Taking the logs of the system constant expressions, we obtain

$$\log [Ag^+] = -11.17 - \log [SCN^-]$$

$$\log [AgSCN] = -7$$

$$\log [Ag(SCN)_2^-] = -3.60 + \log [SCN^-]$$

$$\log [Ag(SCN)_3^{2-}] = -2.09 + 2 \log [SCN^-]$$

$$\log [Ag(SCN)_4^{3-}] = -1.13 + 3 \log [SCN^-]$$

Figure 10-6 shows that the crossed line relation for $[Ag^+]$ vs. $[SCN^-]$ required by simple K_{s0} does occur and the pure water solubility occurs where they cross, because the other complexes are negligible there. The true solubility, however, is the sum of the silver(I) species,

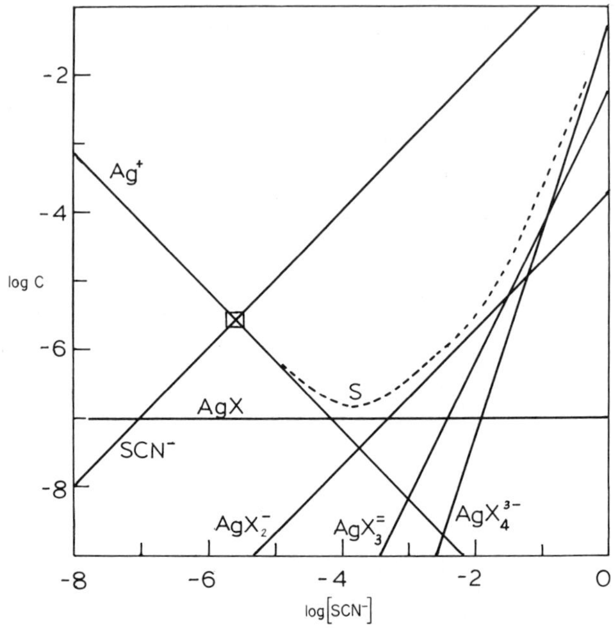

Figure 10-6. Log *C* diagram for saturated solutions of AgSCN in KSCN solutions. *X* is SCN^-. Total solubility is shown dashed.

and, shown by the dashed line, agrees with the experimental curve of Figure 10-4. Above $[SCN^-] = 0.1\ M$, AgSCN dissolves mainly as $Ag(SCN)_4^{3-}$. When such solutions are simply diluted with water, precipitation of AgSCN occurs. The value of the solubility product K_{s0} gives no clue to the solubility in these solutions.

4. Solubilities Dependent upon pH

In the cases considered so far, both ions have been so weakly acidic or basic that we have been able to assume that none of them is removed by reaction with water. Ionic compounds with a basic anion and/or an acid cation are much more common. The carbonates, phosphates, sulfides, fluorides, acetates, and other organic acid anions all take protons in water to reduce the concentration of the anion and increase the solubility of their compounds above that expected from simple K_{s0} behavior.

Example 1. Silver acetate, a monoprotic case. Find the function for solubility as a function of **H** at 25°, $I = 0.1\ M$, with no common ions added, silver acetate as the sole source of silver(I) and acetate ions. This means that $S = C_{\mathrm{M}} = C_{\mathrm{A}}$. The α functions are useful here. We use α_0 as the fraction of ions that are not removed in side reactions, in this case to form AgOH or A^-. With these we write the K_{s0} expression as

$$K_{s0} = C_{\mathrm{M}}\alpha_{0\mathrm{M}} \cdot C_{\mathrm{A}}\alpha_{0\mathrm{A}} = [\mathrm{Ag}^+][\mathrm{A}^-]$$

We know that Ag^+ is not very acidic, so that $\alpha_{0\mathrm{M}} \cong 1$ at moderate p**H** values, and we have the usual $\alpha_{0\mathrm{A}}$ expression for acetate, equation (4-3). The total silver(I) and acetate are S:

$$K_{s0} = S^2(1 + \mathbf{H}/K_a)^{-1}, \qquad S = [K_{s0}(1 + \mathbf{H}/K_a)]^{1/2}$$

This shows the solubility relation to **H**. Putting in the conditional constants $K_{s0} = 10^{-2.16}$ and $K_a = 10^{-4.57}$, we obtain

$$S = (10^{-2.16} + 10^{2.41}\mathbf{H})^{1/2}$$

This function is plotted in Figure 10-7. We see that it does become proportional to p**H** with slope $-\frac{1}{2}$ as **H** becomes large, and as **H** become very small, the basic limit, the solubility levels off at $\sqrt{K_{s0}}$ as expected when $\alpha_{0\mathrm{A}}$ is 1. We have neglected higher silver–acetate

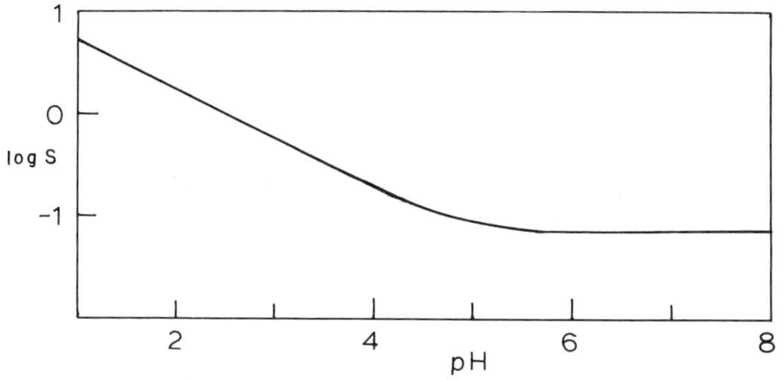

Figure 10-7. The p**H** dependence of the solubility of silver acetate at 25° and 0.1 M ionic strength.

complexing which is important if additional sources of these ions are present.

Example 2. $CaCO_3$. This important mineral substance, calcite, limestone, shell, etc., sets up the following equilibria in water solutions.

The solid dissolves first to ion pairs, or molecules,

$$CaCO_{3(c)} \rightleftharpoons CaCO_3, \qquad K_{s_1} = [CaCO_3] = 10^{-5.20} \quad (10\text{-}5a)$$

These are then involved in dissociation, or, expressed as K_1 (formation),

$$Ca^{2+} + CO_3^{2-} \rightleftharpoons CaCO_3, \qquad K_{1_f} = \frac{[CaCO_3]}{[Ca^{2+}][CO_3^{2-}]} = 10^{2.33} \quad (10\text{-}5b)$$

The quotient of these yields the solubility product:

$$CaCO_{3(c)} \rightleftharpoons Ca^{2+} + CO_3^{2-}, \qquad K_{s_0} = [Ca^{2+}][CO_3^{2-}] = 10^{-7.53} \quad (10\text{-}5c)$$

The Ca^{2+} ion is weakly acidic. This must be considered at high p**H**,

$$Ca^{2+} + H_2O \rightleftharpoons CaOH^+ + H^+_{aq}, \quad K_{a_M} = \frac{\mathbf{H}[CaOH^+]}{[Ca^{2+}]} = 10^{-12.6} \quad (10\text{-}5d)$$

Gas escape of CO_2 must be considered in open systems (considered later),

$$CO_{2(gas)} + H_2O \rightleftharpoons H_2CO_3, \qquad K_P = \frac{[H_2CO_3]}{P_{CO_2}} = 10^{-1.46} \quad (10\text{-}5e)$$

($[H_2CO_3]$ is taken to represent the total dissolved CO_2, much of which may not be truly bonded as carbonic acid molecules.) Finally, the carbonic acid equilibria operate,

$$H_2CO_3 \rightleftharpoons H^+_{aq} + HCO_3^-, \qquad K_1 = \frac{\mathbf{H}[HCO_3^-]}{[H_2CO_3]} = 10^{-6.16} \quad (10\text{-}5f)$$

$$HCO_3^- \rightleftharpoons H^+_{aq} + CO_3^{2-}, \qquad K_2 = \frac{\mathbf{H}[CO_3^{2-}]}{[HCO_3^-]} = 10^{-9.93} \quad (10\text{-}5g)$$

All these set (10-5) constants have been corrected to 0.1 M ionic strength.

It is again useful to express the equilibrium ionic concentrations with $C\alpha$ terms as we did for silver acetate. However, we must stipulate that the C term is the total of only those forms involved in the equilibrium covered by the α and does not include the complex $CaCO_3$ pairs. We have

$$K_{s0} = [Ca^{2+}][CO_3^{2-}] = C_M\alpha_{0M}C_A\alpha_{0A}$$

However, this time, C_M and C_A are equal to S only if $[CaCO_3]$ ion pairs and $[CaOH^+]$ are negligible.

First let us consider the closed system: $CaCO_3$ (solid) + H_2O with no entrance or escape of CO_2 gas. From set (10-5) we see that calcium dissolves to form three species, so that a material balance gives

$$S = [CaCO_3] + [Ca^{2+}] + [CaOH^+]$$
$$S = K_{s_1} + K_{s0}/[CO_3^{2-}] + K_{a_M}K_{s0}/\mathbf{H}[CO_3^{2-}] \qquad (10\text{-}6)$$

The total calcium and carbonate outside the $CaCO_3$ complex are both $S - K_{s_1}$, and the actual $[CO_3^{2-}]$ is $C_A\alpha_{0A}$. We have

$$C_M = C_A = S - K_{s_1} \qquad (10\text{-}7)$$

Using these relations and our α_0 for a diprotic acid, equation (5-4), we obtain

$$S = K_{s_1} + \left[K_{s0}\left(1 + \frac{K_{a_M}}{\mathbf{H}}\right)\frac{\mathbf{H}^2 + K_1\mathbf{H} + K_1K_2}{K_1K_2}\right]^{1/2} \qquad (10\text{-}8)$$

Insertion of the values for the constants for our conditions yields

$$S = 10^{-5.2} + \left[10^{8.56}\left(1 + \frac{10^{-12.6}}{\mathbf{H}}\right)(\mathbf{H}^2 + 10^{-6.16}\mathbf{H} + 10^{-16.09})\right]^{1/2}$$

This solubility function is plotted in Figure 10-8. Below pH 11, the K_{aM} term is negligible, and in the range 10–12 it is small, so that the S function approaches the simple $\sqrt{K_{s0}}$ value, $10^{-3.76}$. In the **pH** range 7–10, most of the dissolved carbonate must form HCO_3^- and the log of S depends on **pH** with slope $-\frac{1}{2}$. Below **pH** 6, the function shifts to dependence on **pH** with slope -1 as the H_2CO_3 term predominates.

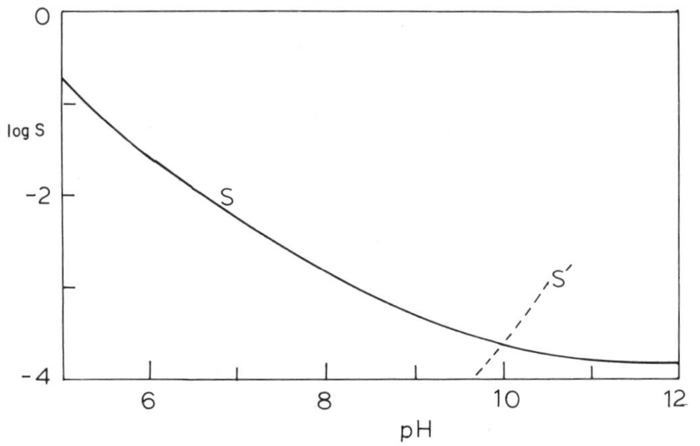

Figure 10-8. The pH dependence of $CaCO_3$ solubility in a closed system (no CO_2 gas equilibrium) at 25° and 0.1 *M* ionic strength.

It is now possible to answer the question, What is the solubility of $CaCO_3$ in pure water at 25° at $I = 0.1\ M$? Both S and **H** are unknown, but they must constitute one point on the curve of Figure 10-8. It must be the point that also answers the question, What is the **pH** of a solution made by dissolving $(S - K_{s_1})$ moles per liter of carbonate ions in water?

It is reasonable to assume that this will be a basic solution and that we can approximate the carbonate equilibrium using only its K_2,

$$CO_3^{2-} + H_2O \rightleftharpoons HCO_3^- + OH^-$$

Let us call the carbonates outside the $CaCO_3$ pairs $y = S - K_{s_1}$ and let $x = [HCO_3^-]$. This also equals the **OH**. In K_2 we get

$$K_2 = 10^{-9.93} = \mathbf{H}(y - x)/x \qquad \text{and} \qquad \mathbf{OH} = x = K_w/\mathbf{H}$$

This produces an equation relating y to **H**,

$$y = (K_w/\mathbf{H})(1 + 10^{-9.93}/\mathbf{H})$$

We can now relate solubility to **H** for these special conditions,

$$S = K_{s_1} + (K_w/\mathbf{H})(1 + 10^{-9.93}/\mathbf{H})$$

A plot of this on the general solubility curve in Figure 10-8 gives the intersection at $10^{-3.61}$ at **pH** 10.0. This shows that our assumption of high **pH** to justify the use of K_2 alone is correct. Elimination of S between this and the general equation (10-8) gives a quartic equation in **H**, so that the graphical method has advantages.

5. Summary

The important points to remember about solubilities are:

1. If a solid dissolves in molecular or ion-pair form, this form has constant activity K_{s_1} in saturated solutions.

2. The solubility of ionic compounds must be calculated from the sum of all dissolved forms. Calculation from K_{s_0} alone yields a minimum solubility correct only if the simple ions alone are the total dissolved form, a rare case.

3. Substitution of equilibrium constant expressions into the material or charge balance equation for solubility can give correct equations for solubility as a function of **H** and/or complexing ligand concentrations.

Problems

1. At 25°, 212 g of sucrose dissolves per 100 g of water. How many water molecules can be associated with each sugar molecule? How many with each —OH group on the sugar? Sugar is $C_{12}H_{22}O_{11}$, F. wt. 342.

2. What is the expected ideal solubility of $BaSO_4$ in pure water; in 0.10 *M* NaCl? $pK^{\circ}_{s_0}$ is 9.96.

3. The experimental solubility of $CsClO_4$ in 0.0334 *M* CsCl was found to be 0.0715 *M*. Calculate K_{s_0} and $K^{\circ}_{s_0}$.

---

4. For the solubility data in the table below, calculate K_s values and comment on the possibilities that simple ionic solubility alone is operating.

Solubilities (*M*) After Saturation with LiF at 30.6 ± 0.2°C[a]

	Pure water	LiF + KF	LiF + LiCl
Li^+	0.0528	0.028	0.113
F^-	0.0528	0.0925	0.025

[a] From C. B. Stubblefield and R. O. Bach, *J. Chem. Eng. Data* **17**, 491 (1972).

5. If equal volumes of 0.2000 *M* RbCl and 0.2000 *M* $HClO_4$ are mixed at 25°, what concentrations are left in solution at equilibrium? Follow the method in this chapter and continue approximations until three significant figures are obtained.

6. pK_{s0}° for $PbSO_4$ at 20° is 7.70. The pK_a for HSO_4^- is 2.00. The K_1° for ion pairs, $PbSO_4$, in the formation direction is 250.
   a. Calculate the simple ideal water solubility, ignoring any ionic interactions.
   b. Derive $K_{s_1}^{\circ}$ and find the water solubility.
   c. Find the solubility in nitric acid solution of **pH** 1.00 and ionic strength 0.10 *M*.

---

7. Show the mathematical details of finding the minimum of the AgSCN solubility curve equation (10-4).

8. Apply the methods used in this chapter to the data of the table below to see if a good fit and intercept can be obtained. Use K_1 for the formation of KSO_4^- pairs as 10.

Solubility of $CaSO_4$ in K_2SO_4 at 25° [a]

K_2SO_4	0 (pure water)	0.0050	0.0100	0.0150	0.0250	0.0565
$CaSO_4$	0.0153	0.0139	0.0128	0.0122	0.0115	0.0107

[a] All values are *M*.

9. Use the results in this chapter for the solubility of silver acetate to construct a log *C* diagram showing logs of acetate, acetic acid molecules, and total solubility.

10. Show that equation (10-4) can be obtained by substituting the α_{0_M} expression for silver ion (equation 8-2) into the K_{s0} expression for AgSCN.

11 Applications of Solubility Calculations

In this chapter, we show calculations of more complex solubility equilibria in chosen cases of importance in geochemistry, biochemistry, and analytical chemistry.

1. $CaCO_3$ Solubility in Open Systems: CO_2 Pressure a Variable

The weathering of carbonate rocks and the precipitation and resolution of $CaCO_3$ in natural waters are problems of intense interest.[1] It is estimated that over 99% of the earth's carbon is present in $CaCO_3$ minerals: limestone, calcite, aragonite, vaterite, marble, etc. To the equilibria of the previous problem, we now add

$$CO_{2(gas)} + H_2O \rightleftharpoons H_2CO_3, \qquad K_P = 10^{-1.46}$$

Here P is expressed in atmospheres. The CO_2 dissolves first to form hydrated molecules, which then slowly rearrange bonds to give carbonic acid molecules. However, as usual, we can be mathematically correct in using an equilibrium constant which relates only the initial and final species as done in K_P [see equation (10-5e)].

Now that carbon can be added or lost from the solution independently of calcium, a material balance will be less useful. The charge balance proves helpful in this case. We consider our system

[1] R. M. GARRELS and C. L. CHRIST, *Solutions, Minerals and Equilibria*, Harper and Row, New York, 1965, Chapter 3.

as made from $CaCO_3$, H_2O, and CO_2 and set equal the totals of each charge species:

$$2[Ca^{2+}] + \mathbf{H} = [HCO_3^-] + 2[CO_3^{2-}] + \mathbf{OH} \qquad (11\text{-}1)$$

With this, and the set (10-5) equilibria, we can relate S to a single variable of interest, the pressure of CO_2 at equilibrium. Let us now derive the expressions we need for each term in (11-1) as a function of S and P. Put the new relation for $[H_2CO_3] = K_P P$ into K_1:

$$K_1 = \mathbf{H}[HCO_3^-]/K_P P$$

Next put it into the product of K_1 and K_2 to get

$$K_1 K_2 = \mathbf{H}^2[CO_3^{2-}]/K_P P$$

Let us neglect $CaOH^+$ (limiting our result to the region below p**H** 12). The total solubility of $CaCO_3$ is

$$S = [Ca^{2+}] + [CaCO_3] \qquad \text{or} \qquad S - K_{s_1} = [Ca^{2+}] = y \qquad (11\text{-}2a)$$

In the K_{s_0} expression, this produces

$$[CO_3^{2-}] = K_{s_0}/y \qquad (11\text{-}2b)$$

Putting this in the $K_1 K_2$ relation, we obtain

$$\mathbf{H} = [K_1 K_2 K_P P y/K_{s_0}]^{1/2} \qquad (11\text{-}2c)$$

which in K_1 gives

$$[HCO_3^-] = [K_1 K_P P K_{s_0}/K_2 y]^{1/2} \qquad (11\text{-}2d)$$

Then,

$$\mathbf{OH} = K_w/\mathbf{H} = K_w(K_{s_0}/K_1 K_2 K_P P y)^{1/2} \qquad (11\text{-}2e)$$

With these expressions we now convert equation (11-1) to

$$2y + (K_1 K_2 K_P P y/K_{s_0})^{1/2}$$
$$= (K_1 K_P P K_{s_0}/K_2 y)^{1/2} + 2K_{s_0}/y + K_w(K_{s_0}/K_1 K_2 K_P P y)^{1/2} \quad (11\text{-}3)$$

Now we can find y at any P, and thus S, which is $y + K_{s_1}$. For our conditions of set (10-5), we insert the K values and multiply equation (11-3) by $y/2$ to get

$$y^2 + 10^{-5.31}P^{1/2}y^{3/2} - (10^{-2.92}P^{1/2} + 10^{-9.29}P^{-1/2})y^{1/2}$$
$$-10^{-7.53} = 0$$

Let us see what effect a CO_2 pressure of 10^{-3} atm has on the solubility compared with that found above in the closed system. This P is a little above normal atmospheric conditions, 0.031 % CO_2. Substituting $P = 10^{-3}$ into the equation just obtained shows that the first and third terms are major ones under these conditions, assuming the S is near 10^{-3} M as obtained in the closed system. We try successive approximations starting with this value, substituting the trial value of S into the $y^{1/2}$ term and solving for a new y from the y^2 term. This rapidly leads to $y = 10^{-2.95}$. This is also S since K_{s_1} is less than 1 % of y. Now we can use set (11-2) to obtain the other species concentrations and find that the **pH** is 7.98. Thus, by the addition of only 10^{-3} atm CO_2, we increase the solubility about fourfold and change the **pH** from about 10 to 8. This shows the powerful effect of CO_2 on the weathering of limestones, and also in the formation of shells by molluscs, who must precipitate $CaCO_3$ from the ions of their tissue fluids.

Ground waters may contain CO_2 pressures much above the open atmospheric condition, due to decay of organic matter. When these waters flow over limestone, some of it dissolves to form channels and caves. Then, if this water laden with CO_2 and dissolved $CaCO_3$ species should drip out of the rocks into the air of a cave, the CO_2 escapes, the **pH** rises, and solid $CaCO_3$ may form stalactites and stalagmites.

A series of CO_2 pressures can be used as just shown to find the solubility and species concentrations as a function of pressure. These are plotted as Figure 11-1. The total solubility lies at, or just above, the Ca^{2+} line of this diagram as in Figure 10-6. Note the parallel lines at the top right side for Ca^{2+} and HCO_3^- just 0.301 unit apart. These suggest the reaction (correct for a certain range)

$$CaCO_{3(c)} + CO_2 + H_2O \rightleftharpoons Ca^{2+} + 2HCO_3^-$$

That is, at high CO_2 pressure and acidity, these must be the major species formed by the dissolving $CaCO_3$. Note that **H** and **OH** are not quite linear functions of this pressure. How can one explain the rise in solubility (Ca^{2+}) at very low CO_2 pressures? If the CO_2 is removed to such a low value as 10^{-8} atm, the LeChatelier shift is in accord with

$$CaCO_3 + H_2O \rightleftharpoons Ca^{2+} + 2OH^- + CO_2$$

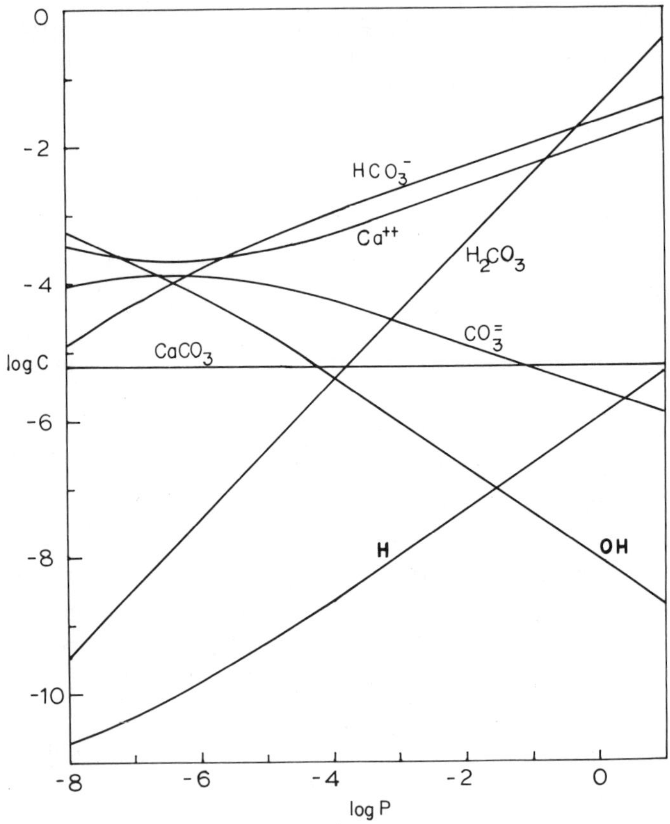

Figure 11-1. The log C diagram for species in saturated solutions of $CaCO_3$ in water under varied pressure of CO_2.

A number of other interesting deductions can be made and correlated with this diagram.

2. Ampholyte Solubility: An Amino Acid

This is another, but simpler, case of the many which are clarified by writing a material balance and reducing it to a function of one variable. Many polyfunctional organic compounds, as well as inorganic compounds having an ampholyte ion, show low solubility in water but become more soluble in both acidic and basic solution. Amino acids and $CaHPO_4$ are examples. Data are available for tyrosine, so we use this as an example to test our equations. It is 2-amino, 3(4-hydroxyphenyl)propanoic acid. Addition of a proton gives the completely protonated form, H_3T^+. The phenolic proton accounts for the third step. A published set of pK_a values for 25° and 0.16 *M* ionic strength is

$$pK_1 = 2.34, \qquad pK_2 = 9.11, \qquad pK_3 = 10.16$$

In addition to the usual three equilibrium steps for a triprotic acid, we have here

$$H_2T_{(c)} \rightleftharpoons H_2T, \qquad K_s = [H_2T]$$

Thus, one species of the triprotic acid is held constant in saturated solutions. This fact can be used in the K_a expressions. For example, in K_1,

$$K_1 = \frac{\mathbf{H}[H_2T]}{[H_3T^+]} = \frac{\mathbf{H}K_s}{[H_3T^+]} \qquad \text{or} \qquad [H_3T^+] = \frac{\mathbf{H}K_s}{K_1}$$

The total solubility must be the material balance sum,

$$S = [H_3T^+] + [H_2T] + [HT^-] + [T^{2-}]$$

Substitution from the four equilibrium expressions above gives

$$S = \mathbf{H}K_s/K_1 + K_s + K_2K_s/\mathbf{H} + K_2K_3K_s/\mathbf{H}^2$$

This same relation results from the obvious condition $K_s = \alpha_2 S$. At high **H**, the first term predominates; at very low **H**, the last two terms become large; and at intermediate **H**, the K_s term is the major one. The experimental solubility of tyrosine in water is 0.0025 *M* and the p**H** found is 5.6. At this distance from the p**H** of pK_1 and pK_2 we know that an α diagram must show α_2 to be over 99% of the tyrosine. (See Chapter 5.) Thus, we take K_s to be 0.0025 *M*. Using this and the other constants in the solubility equation gives the curve in Figure 11-2. Some simple points of interest are:

At p**H** $= pK_1 = 2.34$, $S = 2K_s$.
At p**H** $= pK_2 = 9.11$, $S = 2.01K_s$.
At p**H** from 4 to 7, $S = K_s$ within 1%.

In Figure 11-2, S is expressed in units of K_s, i.e., we give the ratio S/K_s. That is, 2 means S is twice the K_s value. Nine experimental

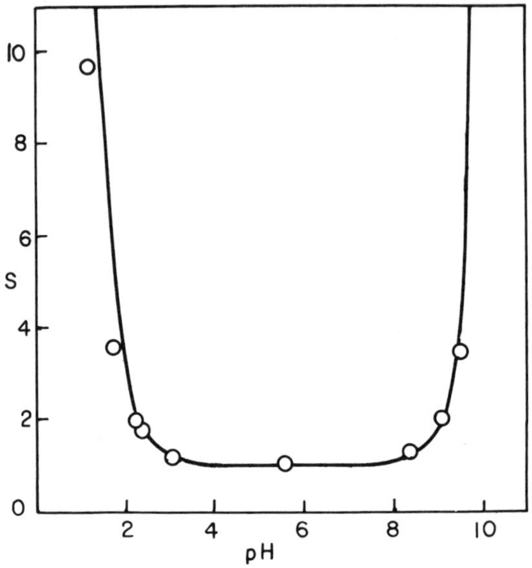

Figure 11-2. The solubility of tyrosine at varied p**H**. Line, predicted; circles, data. S in multiples of S_0, the solubility in water alone.

values are shown as circles. These were reported for 25° but at various and sometimes unstated ionic strengths. The importance of ionic strength was not realized by many workers until recently. The data generally confirm the equation we derived. We can see why it is possible to separate tyrosine from a protein hydrolyzate containing the other, much more soluble amino acids by adjusting the p**H** to the 5–6 range, where it is of low solubility.

If the acidity K values were unknown, the solubility curve would permit evaluation of these constants. The method has not been much used and invites testing.[2]

3. The Solubility of Silver Halides in Ammonia Solutions

The varied solubilities of the chloride, bromide, and iodide of silver(I) have been the basis of many methods for identification and separation. Let us get equations for these solubilities as a function of

[2]G. Anderegg [*Helv. Chim. Acta* **50**, 2333 (1967)] applies this method to EDTA.

unbound $[NH_3]$. We use 25° and a medium 1 M in NH_4NO_3, for which published constants are available. The high concentration of NH_4^+ reduces the **OH** so that its complexing of Ag^+ is small. The solubility sum, material balance on silver(I), is

$$S = [Ag^+] + [AgNH_3^+] + [Ag(NH_3)_2^+]$$

The $\log K_1$ and $\log K_2$ for the complexes are 3.37 and 3.78. We take the halide ion concentration to be S, assuming it to remain unbound under these conditions of very low $[Ag^+]$. Thus, in general for the three halides, we obtain from the K_{so} expression

$$[Ag^+] = K_{so}/S$$

With this and the expressions for the formation constants, we get S as a function of $[NH_3]$ alone:

$$S = K_{so}/S + K_1[NH_3]K_{so}/S + K_1K_2[NH_3]^2K_{so}/S$$

$$S = [K_{so}(1 + K_1[NH_3] + K_1K_2[NH_3]^2)]^{1/2} \qquad (11\text{-}4)$$

As in previous cases, this can also be obtained from α expressions. By definition, $[Ag^+]$ is $\alpha_0 S$, so that $K_{so} = \alpha_0 S^2$.

For AgCl (pK_{so} 9.3), putting in the constants gives

$$S = [10^{-9.3}(1 + 10^{3.37}[NH_3] + 10^{7.15}[NH_3]^2)]^{1/2}$$

Figure 11-3 shows a plot of this equation and those for AgBr (pK_{so} 12.3) and AgI (pK_{so} 16.1). We see that below $[NH_3]$ about 10^{-4} M the solubilities level off at $\sqrt{K_{so}}$. Above about 10^{-3} M NH_3, the diammine term predominates so that S becomes $(K_1K_2K_{so})^{1/2}[NH_3]$, a straight line of slope 1 in the log plot.

Each line so far represents the solubility of the silver halides separately. In mixtures, the silver ion of the more soluble compounds will depress the solubility of the others. We take the ratio of the K_{so} expressions to get the ratio of the solubilities (halide ion ratio) of two compounds, first the chloride to the bromide:

$$\frac{K_{so(Cl)}}{K_{so(Br)}} = \frac{[Ag^+][Cl^-]}{[Ag^+][Br^-]} = 10^{3.0} = \frac{[Cl^-]}{[Br^-]}$$

Thus, the bromide will be three log units below the solubility of the AgCl, shown dashed in Figure 11-3. This assumes that the Ag^+ from the dissolving AgBr is negligible compared with that from the AgCl.

The solubility of silver iodide in the presence of the others, and the solubility of all three halides, can be treated similarly.

The separation factors, defined as the ratio of solubilities, are: Cl^-/Br^-, 10^3; Cl^-/I^-; $10^{6.8}$; Br^-/I^-, $10^{3.8}$. Thus, separations of any pair of halides should be possible with carefully chosen conditions. For example, 100 ml of 1 *M* NH_3 should dissolve about 1.1 g of AgCl, 0.0015 g of AgBr, and 2×10^{-7} g of AgI. Treatment of 1 g of mixed AgCl, AgBr, and AgI with 1 *M* NH_3 can effect the separation of the AgCl with under 1% contamination by the bromide if a large excess volume of the NH_3 is not used. Then treatment with 10 *M* NH_3 could remove AgBr, leaving AgI behind quantitatively. A practical problem is achieving equilibrium with bulk solids in reasonable time. Quan-

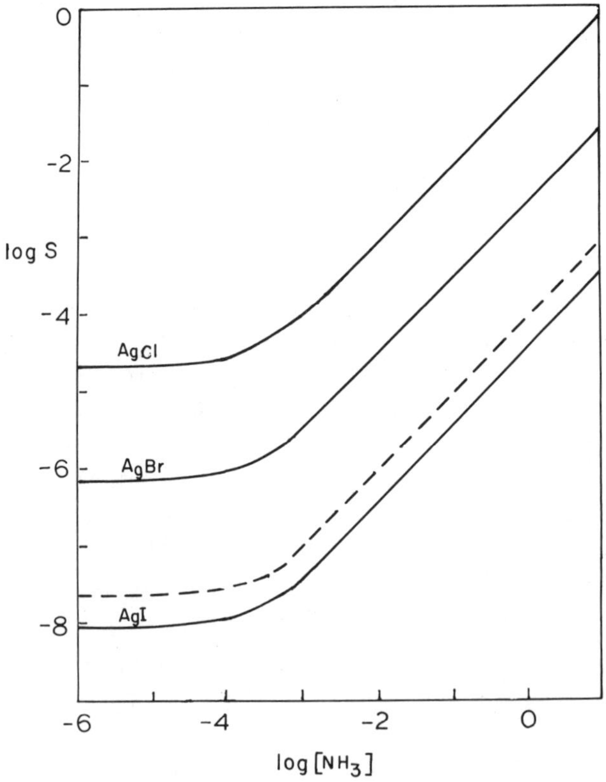

Figure 11-3. The solubilities of silver halides in NH_3 solutions: solid lines, alone; dashed line, AgBr in saturated AgCl.

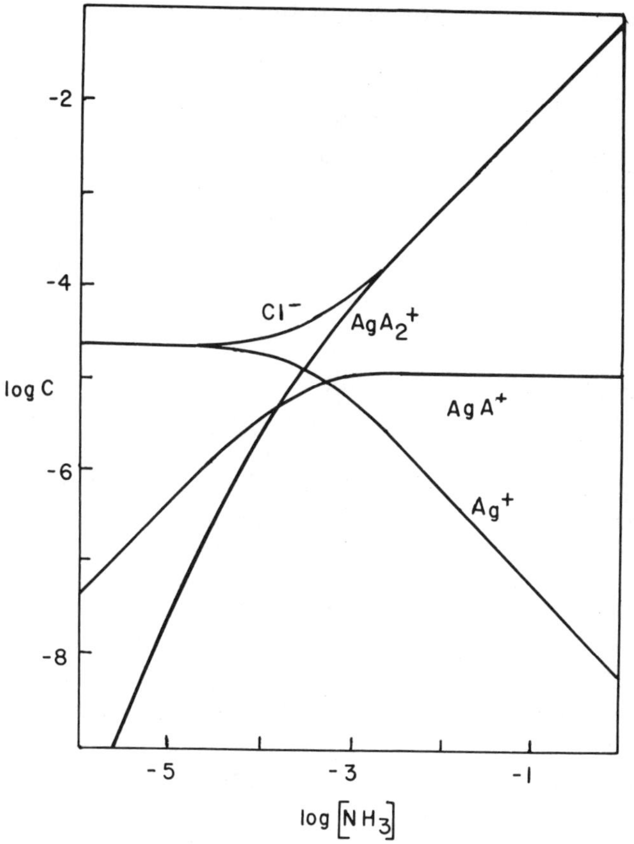

Figure 11-4. The log C diagram of species in saturated AgCl solutions with varied NH_3 concentrations. A is NH_3.

titative separation methods have been devised more easily in the reverse direction. Slow evaporation of the NH_3 from solutions containing the mixed halides permits the more insoluble to precipitate and be filtered out before the next one in solubility. Clearly, from the K_{so} value, very little I^- can be present. This is an example of homogeneous precipitation.

The log C diagram can be constructed just as was done for AgSCN and $CaCO_3$. It is shown as Figure 11-4 for AgCl. The solubility (Cl^-) line is the same as that in Figure 11-3. It is instructive to compare this with the α diagram for silver(I)–NH_3 (Figure 9-1).

4. Mineral Precipitation from Sea Water

Analytical concentrations of materials in sea water are listed in Chapter 9, Section 6. They suggest that the following common mineral precipitates might be supersaturated: $CaCO_3$ (calcite), $MgCO_3$ (magnesite), $CaMg(CO_3)_2$ (dolomite), and $CaSO_4$ (gypsum). The ionic activities used in Chapter 9 are ($f \times M$)

$$(Ca^{2+}) = 2.5 \times 10^{-3}, \qquad (CO_3^{2-}) = 4.9 \times 10^{-6}$$

$$(Mg^{2+}) = 1.7 \times 10^{-2}, \qquad (SO_4^{2-}) = 1.8 \times 10^{-3}$$

A comparison of the 25° K_{so}° values with the appropriate products of these activities gives the results shown in Table 11-1. Thus, even if activity is taken into account, it seems that only $CaSO_4$ should remain unprecipitated in the seas. What are we to make of this? Supersaturation seems unlikely since there is sufficient time, agitation, and contact with solids to bring about equilibrium. Some explanations that have been made are:

1. Activity coefficients and formation constants for the complexes (Chapter 9) contain enough uncertainty to account for the discrepancy and the sea is actually just near saturation in these carbonates.
2. Organic matter in the sea complexes, and lowers further, the unbound Ca^{2+} and Mg^{2+}.
3. The sea really is supersaturated in these compounds and some unknown phenomenon prevents the crystallization of the solids.

None of these is particularly convincing and chemical oceanographers are actively seeking better answers. The concentration of dissolved CO_2 at various depths was measured around the world and the solubility of $CaCO_3$ recalculated with new constants obtained

Table 11-1

	$CaCO_3$	$MgCO_3$	$CaMg(CO_3)_2$	$CaSO_4$
Published pK_{so}°	−8.35	−7.5	−16.7	−4.6
Sea ion product	−7.9	−7.1	−15	−5.3

Table 11-2

$CaCO_3$ form	Crystal structure	Density	K_s° (25°)
Calcite	Rhombohedral	2.71	4.5×10^{-9}
Aragonite	Rhombic	2.93	6.0×10^{-9}
Vaterite	Hexagonal	2.64	—

empirically at the high ionic strength of sea water.[3] In this way, assumptions about the activity coefficients are not required. These workers point out that $CaCO_3$ becomes less soluble with temperature rise, and more soluble with increased pressure deep in the seas. They find that $CaCO_3$ is forming on the shallow continental shelves in tropical seas and that very little is falling to the ocean depths. They were still forced to conclude that the surface waters are slightly supersaturated in $CaCO_3$. It is definitely undersaturated at depths of 5000 m.

5. The Problem of Uniqueness of the Solid Phase

In experimental study of solubilities, it is important to know the nature of the solids precipitated. Lack of agreement in K_{so} among different mixtures can be the result of slow changes of one crystal form to another or of formation of another solid of different composition. Calcium carbonate can precipitate in two forms under different conditions and a third rare form, vaterite, is known (Table 11-2). Calcite and aragonite are well known in nature, although calcite is the less soluble and, thus, favored form at equilibrium. We have assumed it to be the solid in all our calculations on $CaCO_3$.

Calcium sulfate precipitates from water as gypsum, the dihydrate, $CaSO_4 \cdot 2H_2O$. If water can be taken to be at unit activity, or at least as constant, the mathematical form of our treatment is not changed.

However, many published K_{so} values refer to fictitious or poorly characterized solid phases. "$Fe(OH)_3$" and most metal hydroxides

[3] Y.-H. Li. T. Takahashi, and W. Broecker, *J. Geophys. Res.* **74**, 5507 (1969); J. H. Carpenter, "Problems in Applications of Analytical Chemistry to Oceanography," in *Analytical Chemistry: Key to Progress on National Problems*, NBS Special Publication 351, Washington, D. C., 1972, Chapter 7 (a review with 33 references).

are in this class. Iron(III)$_{aq}$ polymerizes as the **pH** rises and then precipitates as hydrous oxides ranging in composition through $Fe(OH)_3$, $FeO(OH)$, to Fe_2O_3. Thus, a reported K_{s0} is just a guide to the **pH** at which precipitation may be expected under some conditions.

For valid solubility investigation, one must verify the composition and crystal form of precipitates or initial solids. In the bone and rock phosphate discussion following, the $CaHPO_4$ is a definite crystalline compound, while apatite and $Ca_3(PO_4)_2$ do not seem to precipitate from water in reproducible forms. Basic phosphates with a range of composition seem to occur, according to some workers in the field. We use the K_{s0} values reported as guides to precipitation behavior only.

6. The Calcium–Phosphate System: Bone and Mineral Formation

Human bone contains 30–35% organic matter, mainly the protein collagen, and 65–70% inorganic matter, dry basis. The inorganic matter is largely very small crystals with a lattice structure and composition approximately that of apatite, $Ca_5(PO_4)_3OH$. Substitution of ions occurs: equivalent numbers of Mg^{2+}, K^+, or Na^+ may replace the Ca^{2+}, and a Cl^-, F^-, or HCO_3^- sometimes fills the OH^- position in the lattice. All these apatites are common in nature, and used for sources of phosphorus chemicals and fertilizer. In bone, the large exposed surface areas of the small crystals and their hydration help account for the ease of dissolution and reprecipitation of the calcium phosphates during bone growth and/or resorption.[4] Here we examine the solubility relations of various proposed bone minerals. Because of the lack of precise data and characterization of the calcium phosphates, this treatment should be read as the method of approach rather than as a complete quantitative description. For 25° and $I = 0.1\ M$, and phosphate system values in Chapter 5,

[4] G. Bell, J. Davidson, and H. Scarborough, *Textbook of Physiological Biochemistry*, 6th ed., Williams and Wilkins, Baltimore, Maryland, 1965, Chapter 15.

we have

$$\text{CaHPO}_4, \qquad K_{so}^\circ = (\text{Ca}^{2+})(\text{HPO}_4^{2-}) = 10^{-7.0} \qquad (11\text{-}5a)$$

$$\text{Ca}_5(\text{PO}_4)_3\text{OH}, \qquad K_{so}^\circ = (\text{Ca}^{2+})^5(\text{PO}_4^{3-})^3\textbf{OH} = 10^{-57.8} \qquad (11\text{-}5b)$$

$$\text{Ca}_3(\text{PO}_4)_2, \qquad K_{so}^\circ = (\text{Ca}^{2+})^3(\text{PO}_4^{3-})^2 = 10^{-26.0} \qquad (11\text{-}5c)$$

$$\text{ion pairs,} \qquad K_{f_1}^\circ = \frac{(\text{CaHPO}_4)}{(\text{Ca}^{2+})(\text{HPO}_4^{2-})} = 10^{2.70} \qquad (11\text{-}5d)$$

$$K_{f_2}^\circ = \frac{(\text{CaH}_2\text{PO}_4^+)}{(\text{Ca}^{2+})(\text{H}_2\text{PO}_4^-)} = 10^{1.08} = 12 \qquad (11\text{-}5e)$$

As we found before for the first type of ion pair, the product of equations (11-5a) and (11-5d) gives a constant value for the molecules in saturated $CaHPO_4$ solutions:

$$K_{so}^\circ K_{f_1}^\circ = (\text{CaHPO}_4) = 10^{-4.30} = K_{s_1} = 5 \times 10^{-5}\ M$$

We shall take the activity equal to the molarity for this uncharged species.

Normal serum calcium level in man is 4.5–5.5 $\times 10^{-3}$ *M*. About 40% of this is bound to protein (nondialyzable) and is not in rapid equilibrium with the solution. Thus, we use about 3×10^{-3} *M* for the available calcium(II). Similarly, there is about 1×10^{-3} *M* available phosphates, mainly $H_2PO_4^-$ and HPO_4^{2-} (see α diagrams). This is higher in children during bone-forming years. First, we might ask whether these levels are near or above saturation for the solids listed above. To find the appropriate phosphate ion concentration, we need only multiply the total available phosphates by the α value for blood serum pH, 7.40. We use the Kielland table (Appendix A-1) values of activity coefficients to adjust the K values to $I = 0.1\ M$. The f values are: for Ca^{2+}, 0.405; for PO_4^{3-}, 0.095; for HPO_4^{2-}, 0.37; for OH^-, 0.755. This produces new conditional K values for the set (11-5a)–(11-5e):

(a) $10^{-6.18}$, (b) $10^{-52.6}$, (c) $10^{-22.8}$, (e) 4.85

Using the phosphate system $\alpha_0 = 4 \times 10^{-5}$ and $\alpha_1 = 0.80$, we can test the serum values in the K expressions to see what the actual ion products seem to be.

(a) $[Ca^{2+}][HPO_4^{2-}] = (3 \times 10^{-3})(10^{-3})0.80 = 2._4 \times 10^{-6}$

This is somewhat above the conditional K_{s0}. However, published values range over ± 0.5 log units and the effect of temperature is not clear (body at 37°C). Thus, all we should say is that the serum is at, or near, saturation in $CaHPO_4$.

(c) $Ca_3(PO_4)_2$, $[Ca^{2+}]^3[PO_4^{3-}]^2 = (3 \times 10^{-3})^3(10^{-3})^2(4 \times 10^{-5})^2$

$$= 4 \times 10^{-23}$$

which is close to the conditional K_{s0} of 1.6×10^{-23}.

(b) $Ca_5(PO_4)_3OH$,

$$[Ca^{2+}]^5[PO_4^{3-}]^3\mathbf{OH} = (3 \times 10^{-3})^5(10^{-3})^3(4 \times 10^{-5})^3(10^{-6.4}) = 10^{-41.2}$$

(13.8 was used for pK_w at this ionic strength.) This is considerably above the saturation value of $10^{-52.6}$. Thus, we may conclude that serum is most clearly supersaturated with respect to apatite at blood pH. This may become clearer by looking at the variation of these solubilities with **pH**, and bringing in the effects of complexing (11-5d) and (11-5e). Cells may alter **pH** locally to make bone.

(*i*) *CaHPO₄*. We do a material balance on total dissolved calcium(II):

$$\text{total}[Ca(II)] = S = [Ca^{2+}] + [CaHPO_4] + [CaH_2PO_4^+]$$

Substitute from the K expressions to obtain functions of **H** alone:

$$S = (K_{s0}/\alpha_1)^{1/2} + K_{s_1} + K_{f_2}K_{s0}\alpha_2/\alpha_1$$

Note that the Ca^{2+} is the calcium(II) not in the ion pair complexes and this equals the phosphates in protonic equilibria, not in Ca complexes. This allows the substitutions made above. Next, we find the α values from the curves and calculate S at chosen **pH** values. We repeat this process for the other two solids to obtain the curves of Figure 11-5.

(*ii*) For apatite, the total dissolved calcium(II) is five times the molar solubility of $Ca_5(PO_4)_3OH$. Thus, the material balance on Ca is

$$\text{total}[Ca(II)] = 5S = [Ca^{2+}] + [CaHPO_4] + [CaH_2PO_4^+]$$

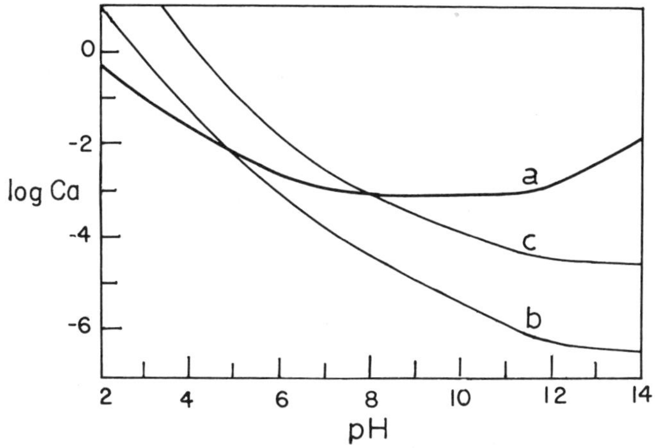

Figure 11-5. Solubilities of possible calcium phosphates at varied **pH**. (a) $CaHPO_4$, (b) apatite, $Ca_5(PO_4)_3OH$, (c) $Ca_3(PO_4)_2$.

This time, we cannot rigorously assume that

$$[Ca^{2+}]/5 = [\text{protonated phosphates}]/3$$

since the two calcium phosphate complexes remove the species in 1:1 proportions, not 5:3. However, for a first numerical approximation let us use the 5:3 ratio starting with K_{s_0} as before. We have

$$K_{s_0} = [Ca^{2+}]^5[PO_4^{3-}]^3\mathbf{OH} = (5C)^5(3C\alpha_0)^3\mathbf{OH} = 5^5 3^3 C^8 \alpha_0^3 K_w/\mathbf{H}$$

$$C = [K_{s_0}\mathbf{H}/5^5 3^3 \alpha_0^3 K_w]^{1/8}$$

where C is the molarity of dissolved apatite which is in equilibrium with the Ca^{2+} and protonated phosphates. $[Ca^{2+}] = 5C$ and the [total protonated phosphates] $= 3C$.

We choose **pH** values, calculate C, and then use α_1 and α_2 to estimate $[CaHPO_4]$ and $[CaH_2PO_4^+]$ from equations (11-5d) and (11.5e). The $[CaHPO_4]$ proves negligible (assuming no solid $CaHPO_4$ is present). Using the conditional constant 4.85, we calculate the second complex with equation (11.5e),

$$[CaH_2PO_4^+] = 4.85(5C)(3C\alpha_2)$$

Below **pH** 4, this becomes rather large and the approximations fail. However, the trends of the curve for apatite, curve b in Figure 11-5, are correct.

(*iii*) Finally, for tricalcium phosphate, $[CaHPO_4]$ again is negligible. The calcium balance gives

$$\text{total}[\text{Ca(II)}] = 3S = [\text{Ca}^{2+}] + [\text{CaH}_2\text{PO}_4^+]$$

As above, we use K_{so} to find the ionic calcium and phosphates. We have

$$[\text{Ca}^{2+}]^3[\text{PO}_4^{3-}]^2 = K_{so} = (3C)^3(2C\alpha_0)^2, \qquad C = [K_{so}/3^3 2^2 \alpha_0^2]^{1/5}$$

We take Ca^{2+} as $3C$, and use the same method for $CaH_2PO_4^+$,

$$[\text{CaH}_2\text{PO}_4^+] = 4.85(3C)(2C\alpha_2)$$

This produces curve *c* in Figure 11-5.

Thus, we can now see why apatite forms preferentially above p**H** 5 in the body and why the laboratory study of $CaHPO_4$ solubility succeeds only at p**H** below 5, where the other solids do not form. The rise of $CaHPO_4$ solubility at higher p**H** is typical of ampholyte materials, as we saw in the case of tyrosine.

Bone formation is clearly not a simple precipitation of apatite (see reference in footnote 4). A complex set of enzymatic cell reactions controls it. However, we can say that the most stable calcium phosphate is used and that the system at p**H** 7.4 is operating very close to equilibrium. Furthermore, it is plausible to find apatite the most abundant of the phosphate minerals in nature, where waters range in p**H** from about 5–9.

7. Solubility Affected by Amino Acid Chelation

High-precision and high-accuracy determination of amino acids is difficult. Let us examine the possibilities of an indirect approach through dissolving a metal ion compound followed by the highly accurate EDTA (or other) metal ion methods. An iodide–Cu(II) method was proposed in 1950.[5] Likely candidates are Cu(II) and Hg(II), which form very stable complexes with amino acids. The glycinate complexes have the stability constants shown in Table 11-3. The objective is to get quantitative dissolving of a low solubility metal ion compound when excess of it is stirred with a measured

[5] W. A. Schroeder, L. M. Kay, and R. S. Mills, *Anal. Chem.* **22**, 760 (1950).

Table 11-3

	$\log K_1$	$\log K_2$	Conditions
Cu^{2+}	8.07	6.90	25°, 0.1 M KNO_3
Hg^{2+}	10.3	8.9	20°, 0.5 M KNO_3

sample of the amino acid solution. Clearly, the metal compound must have low solubility to keep a small blank, but enough solubility to allow quantitative amino acid complexing. Take the simple compound of a -2 anion, like $CuCO_3$,

$$CuA_{(c)} + 2Gly^- \rightleftharpoons Cu(Gly)_2 + A^{2-}$$

Looking at the constant expressions, we see that the equilibrium constant for this overall reaction is the product

$$K_{eq} = K_1 K_2 K_{so} = \frac{[CuGly_2][A^{2-}]}{[Gly^-]^2}$$

For $CuCO_3$, K_{so} is about 10^{-10}, so that this K_{eq} is about 10^5. Putting a complete reaction of C moles per liter of glycinate into this at high **pH** (about 9), where α_{0c} (CO_3^{2-}) is about 0.01 and α_{0g} (glycine) is about 0.1, we get

$$K_{eq} = 10^5 = \frac{(C/2)^2\alpha_{0c}}{(x\alpha_{0g})^2}$$

where x is the small amount of glycinate left in solution unbound. Putting in the α values, we get the ratio

$$C/x = 10^3$$

Thus, the reaction seems barely useful. Furthermore, HCO_3^- complexes Cu(II) and greatly increases its solubility under some conditions. A suggestion which does work is to use a copper(II) phosphate (see footnote 5). A K_{so} for this compound has been published as 10^{-37}. One may well doubt the existence of a pure solid precipitate of this composition for reasons discussed previously, but we take this constant as an indication of solubility at least to see if the method is rational. The reaction we visualize is

$$Cu_3(PO_4)_{2(c)} + 6Gly^- \rightleftharpoons 3CuGly_2 + 2PO_4^{3-}$$

Putting together the constant expressions as usual, we get

$$K_{eq} = (K_1K_2)^3K_{so} = 10^8 = \frac{[CuGly_2]^3[PO_4^{3-}]^2}{[Gly^-]^6}$$

If C moles per liter of glycinate react, we get $(C/2)CuGly_2$ and $(C/3)PO_4^{3-}$ (leaving x glycinate), so that

$$K_{eq} = \frac{(C/2)^3(\alpha_{0p}C/3)^2}{x^6} = \frac{C^5\alpha_0^2}{72x^6}$$

This is done in a borax buffer at **pH** 9, where phosphate α_0 is 2×10^{-3}. This gives us the ratio

$$C^5/x^6 = 2 \times 10^{15}$$

Here we took α_0 for glycine as about 1. At least the reaction is favored, but the quantitative quality is not assured. At $C = 0.01\ M$ we calculate x as $10^{-4}\ M$, about 1%, which is not promising for high (four figure) accuracy. This is hardly a complete treatment. We must ask if $Cu(OH)_2$ or the borate may not also be solid forms, and what is the relative effect of **pH** on the glycinate and phosphate, or other anions. A similar process for mercury using the constant 10^{-25} for

$$HgO + H_2O \rightleftharpoons Hg^{2+} + 2OH^-$$

shows that the Hg(II) complexing case will not be as favorable as the Cu(II) above. The α_0 for glycine becomes less favorable as the solubility increases at lower **pH**.

Problems

1. Use $K_{so} = 10^{-16.1}$ for AgI and $\log K_1$ and $\log K_2$ of 9.0 and 4.0 for the system Ag^+–$S_2O_3^{2-}$ and derive a relation for the solubility of AgI in thiosulfate solutions. Assume all the dissolved silver is in complexes with thiosulfate. Sketch a diagram of $\log S$ and $\log C$ of the complexes vs. $S_2O_3^{2-}$.

2. Compare the solubility of AgCl, AgBr, and AgI in 1 M $S_2O_3^{2-}$ solution, about that used in the photographic removal of these compounds from film. (See previous problem.)

3. Approach this by analogy with $CaCO_3$ in the chapter. The $\log K_P$ for H_2S at 25° is −1.0 and K_1K_2 for H_2S is $10^{-20.0}$. Find these solubility functions for ZnS

assuming $S = Zn^{2+}$ and no other soluble form of Zn(II) is formed. The log K_{so} for ZnS is -23.8. Find:

a. The solubility as a function of **H** at constant H_2S pressure of 1 atm.
b. The solubility of ZnS as a function of P_{H_2S} in a buffer at constant **pH** of 9.0.
c. The solubility of ZnS in water as a function of P_{H_2S} allowing **H** to adjust to what is required by S and P.

4. At 1 atm H_2S, and in 0.010 M Zn^{2+}, what **H** will just keep ZnS from precipitating? See previous problem for data.

---

5. HgS is one of the lowest solubility sulfides, log $K_{so} = -52.7$. Compare the amounts dissolved in 1–10 M $S_2O_3^{2-}$ and I^- at **pH** 0. The β_2 for Hg(II)–$S_2O_3^{2-}$ is 10^{29}, and β_4 for HgI_4^{2-} is 10^{30}. Simplify by writing an overall equation for dissolving HgS to give H_2S and the highest complex.

6. How many grams of $CaCO_3$ precipitate and how many milliliters of CO_2 at STP evolve when one liter of saturated $CaCO_3$ solution at $P_{CO_2} = 0.10$ atm is brought to a new equilibrium with P_{CO_2} 10^{-3} atm? Estimate by using Figure 11-1.

7. Find the solubility of $CaHPO_4$ at **pH** 3.0 and at **pH** 7.0. Point out how the various terms in the equation act to make the solubility at **pH** 3 about 100 times that at **pH** 7.

8. When $Cu_3(PO_4)_2$ is brought to equilibrium with 0.0200 M glycine solution at **pH** 9.00, calculate the concentrations of the Cu(II) species and see if one is justified in using $\bar{n} = 2.00$ for glycine complexes of Cu(II). See data in this chapter.

9. Look at the conditions we decided were valid in Chapter 10 for the solubility and **pH** of $CaCO_3$ in water in a closed system and find this point on Figure 11-1. Do the species values agree with those calculated in Chapter 10? The P_{CO_2} cannot be zero. What is it?

12 Oxidation–Reduction

1. Definitions and Relations of Electrochemistry

Redox reactions are usually described by a disguised form of equilibrium constant called the standard cell potential. It is directly proportional to the log of the K°_{eq}. A system of hypothetical half-cell potentials is based on the standard hydrogen electrode. The Nernst equation relates cell potentials to the log of activities involved and to the K°_{eq}. For a reaction to which a net change in oxidation states n applies,

$$aA + bB + \cdots \rightleftharpoons xX + yY + \cdots$$

$$E = E^\circ - \frac{0.05916}{n} \log \frac{(X)^x(Y)^y \cdots}{(A)^a(B)^b \cdots} \qquad (12\text{-}1)$$

When all the activities are unity (or some combination of them makes the quotient 1), then $E = E^\circ$.

When equilibrium activities are reached, there is no net driving free energy for reaction and E reaches zero. At equilibrium, at 25°, this gives

$$E^\circ = \frac{0.05916}{n} \log K^\circ_{eq} \qquad (12\text{-}2)$$

This comes about because the quotient in the log expression in equation (12-1) has the form of the equilibrium constant expression. But it is K°_{eq} only when equilibrium is reached. At other conditions its log measures the free energy (cell potential) drive toward equilibrium. Calling the quotient Q, we can write

$$E = E^\circ - \frac{0.05916}{n} \log Q = \frac{0.05916}{n} (\log K^\circ_{eq} - \log Q)$$

This stresses that the measured cell potential E depends on the difference between equilibrium conditions and the given Q conditions. Other relations of use in thermodynamic calculations relate these to the standard free energy change:

$$\Delta G^\circ = -RT \ln K^\circ_{eq} = -nFE^\circ$$

The constant 0.05916 V-equiv/mol is $2.303RT/F$, where R is the gas constant, $T = 298.16°K$, and F is the faraday.

Redox reactions are often divided, in thought, into oxidation and reduction half-reactions. A table of these may be devised by relating them to any one standard. Hydrogen is used. For example, for silver the reaction imagined, and which can be measured experimentally, is

$$Ag^+ + \tfrac{1}{2}H_2 \rightleftharpoons H^+_{aq} + Ag^\circ$$

The Nernst expression for its cell potential is

$$E = E^\circ - 0.05916 \log \frac{(H^+)(Ag^\circ)}{(Ag^+)(H_2)^{1/2}}$$

The potential of the standard hydrogen half-cell can be set at zero since only differences are measured, and the hydrogen value always cancels in these differences. In the cell above, the activities of H_2, H^+_{aq}, and Ag° are 1, as defined for the standard hydrogen half-cell and a pure solid. Putting these into the Nernst expression above shows why we can call this the half-cell potential for the silver couple:

$$E = E^\circ_{1/2} - 0.05916 \log (Ag^+)^{-1} \qquad (12\text{-}3)$$

for the imagined half-reaction

$$Ag^+ + e^- \rightleftharpoons Ag^\circ$$

But, of course, no half-cell has been measured. The complete cell versus standard hydrogen is implied in half-cell potentials. The symbols E_H or E_h are sometimes used to indicate the basis on the hydrogen scale. This is wise in practice because actual laboratory measurements are often made against more convenient mercury or silver half-cells, whose value must then be used to convert a cell potential to the hydrogen basis. A further caution is that the electrical sign of the electrode must be determined before the potentials of the two halves of a cell can be deduced. That is, an absolute voltage on a cell of 0.3 V might be the result of the known half at 0.4 V and an unknown of either 0.1 or 0.7 V, depending on sign.

We shall follow the practice of denoting all standard potentials in the reduction direction, called the IUPAC, or Stockholm convention. This has the advantage, over the oxidation potential method, of giving the potentials the electrostatic sign of the electrode when connected to standard hydrogen, and also the mathematical sign indicating spontaneity (+) or nonspontaneity (−) in the Nernst equation. The methods are evaluated in detail by Bockris and Reddy.[1]

In the case of the silver couple, $E^\circ_{1/2}$ is $+0.7991$ V, which tells us that it is positive with respect to hydrogen: it pulls electrons from H_2 as Ag^+ goes to Ag°. Furthermore, it tells us that this reduction direction is spontaneous versus hydrogen. The K°_{eq} calculated from equation (12-2) is $10^{13.51}$, showing that the reaction must go far to the right to reach equilibrium starting from unit activities.

The reverse method, oxidation potentials (the American convention) is seen in many books and is used by some practicing chemists in their publications. Vigilance is required to deduce what is meant by "electrode potential" without the qualifying terms, reduction, oxidation, or IUPAC convention, etc. Careful attention to directions and signs, holding to one consistent form of the Nernst equation, is recommended to avoid sign mistakes in calculations.

2. Log Activity versus Potential Diagrams

These simple diagrams may help in understanding the more subtle E–pH diagrams which have wider uses, but which are more difficult to construct.

[1] J. O. Bockris and A. K. N. Reddy, *Modern Electrochemistry*, Plenum Press, New York, 1970, Vol. II, pp. 1115–1120.

Equation (12-3) is plotted in Figure 12-1 for the case 0.1 mol (10.787 g) silver in contact with one liter of solution. Silver metal, at unit activity (remember this is independent of amount of silver) is present until the (Ag^+) reaches the 0.1 *M* value. Since the estimated Debye–Hückel activity coefficient is 0.75, this is

$$(Ag^+) = 0.10(0.75) = 0.075, \qquad \log(0.075) = -1.125$$

This occurs when sufficient oxidizing agent (unspecified) is added to raise the *E* to [equation (12-3)]

$$E = 0.7991 - 0.05916(1.125) = 0.733 \text{ V}$$

The (Ag^+) increases to this point as a straight line of slope 0.05916 V per logarithmic activity unit, intercepting $\log a = 0$ at $E = 0.7991$ V.

A similar treatment for iron reveals the behavior of a system having two soluble oxidation states (Figure 12-2). The two reactions being considered are

$$Fe^{2+} + 2e^- \rightleftharpoons Fe^\circ, \qquad E^\circ = -0.440 \text{ V} \qquad (12\text{-}4)$$

$$Fe^{3+} + e^- \rightleftharpoons Fe^{2+}, \qquad E^\circ = 0.771 \text{ V} \qquad (12\text{-}5)$$

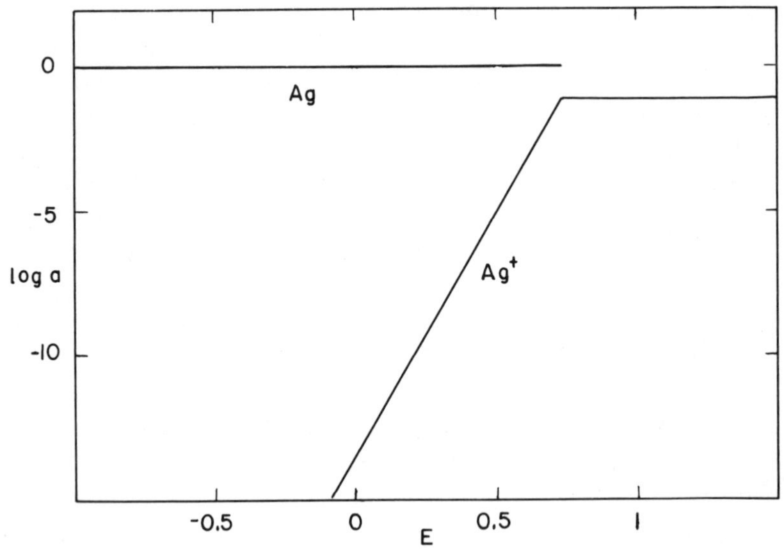

Figure 12-1. Log activity versus *E* in volts for the system of 0.100 mol Ag per liter at 25° and 0.1 *M* ionic strength.

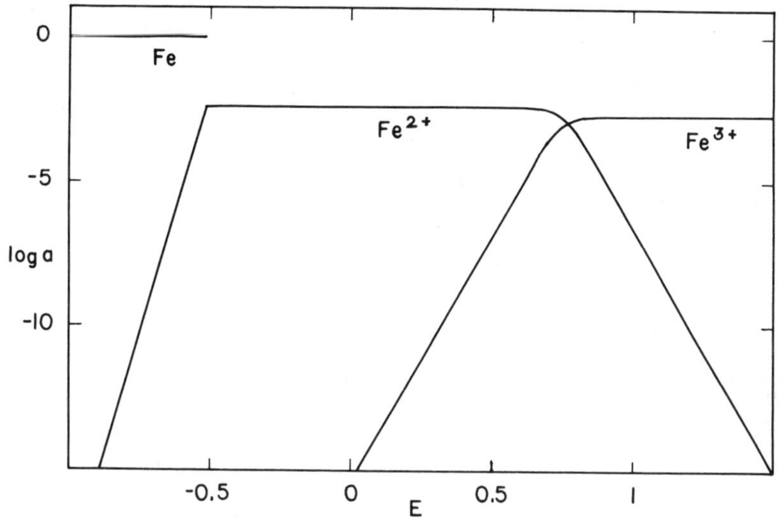

Figure 12-2. Log activity versus E in volts for the system of 0.0100 mol Fe per liter at 25° and 0.1 M ionic strength.

The corresponding Nernst equations are

$$E = -0.440 - (0.05916/2) \log (Fe^{2+})^{-1} \quad (12\text{-}6)$$

$$E = 0.771 - 0.05916 \log [(Fe^{2+})/(Fe^{3+})] \quad (12\text{-}7)$$

This time, let us take the total dissolved iron as 0.01 M and ionic strength 0.1 M (this comes about because of the higher charges, the z^2 factor, and because some acid will be required in practice to prevent $FeOH^{2+}$ and other basic ion formation). Estimated iron(II) and iron(III) activity coefficients are 0.405 and 0.18 (Kielland table Appendix A-1). So, the activities of each at their maxima are

$$(Fe^{2+}) = 0.01(0.405) = 4.0 \times 10^{-3}, \qquad \log a = -2.40$$

$$(Fe^{3+}) = 0.01(0.18) = 1.8 \times 10^{-3}, \qquad \log a = -2.74$$

We use equation (12-6) to get the first (left) portion of the curve just as we did for the silver case. (Compare Figure 12-1.) The slope is greater because of the different n factor. The E at which $\log (Fe^{2+})$ becomes -2.40 is -0.51 V. All the iron metal has dissolved and the Fe(II) remains constant until the E value becomes high enough to convert significant amounts of it to Fe(III).

Let us point out the ways to calculate the activities in these solutions, since certain new situations arise in the region having iron(II) and (III) present in significant amounts. It is simple to choose values of E and solve for the activity ratio in equation (12-7). At $E = 0.5$ V,

$$0.500 = 0.771 - 0.05916 \log R$$

$$R = 10^{4.58} = (Fe^{2+})/(Fe^{3+})$$

Since the iron is quantitatively in the (II) form at so high a ratio, we can take (Fe^{2+}) as still $10^{-2.40}$ and solve for (Fe^{3+}) as $10^{-6.98}$. A similar calculation at 1.00 V, where (Fe^{3+}) is now $10^{-2.74}$, gives $(Fe^{2+}) = 10^{-6.61}$. A few more points establish the lines in Figure 12-2.

In the curved portions of the lines, a different approach is required. From the Nernst equation (12-7), we see that the iron(II) and (III) become equal in activity at 0.771 V. We can set equal the molarity times activity coefficient for each, since the total molarity is 0.01 M. Let x be the molarity of iron(II):

$$x(0.405) = (0.01 - x)0.18$$

This gives us the molarities, $[Fe^{2+}] = 0.00308$ and $[Fe^{3+}] = 0.00692$. The activity of each ion is then $10^{-2.906}$ at 0.771 V.

The resemblance of this ratio to a monoprotic acid–base ratio can be seen by comparison of equation (12-7) with

$$\text{pH} = \text{p}K_a - \log([\text{HA}]/[\text{A}])$$

where the activity coefficients are concealed in the effective pK_a. We can achieve the same form here by getting a corrected ratio for iron(II) to (III), following the previous paragraph,

$$R = x(0.405)/(0.01 - x)0.18, \qquad R(0.444) = x/(0.01 - x)$$

The alpha fractions for iron(II) and iron(III) can now be obtained from the activity ratios at selected E values in equation (12-7), since

$$\alpha_{\text{II}} = x/0.01 \qquad \text{and} \qquad \alpha_{\text{III}} = (0.01 - x)/0.01$$

This has been plotted in Figure 12-3. Note that the activities of the soluble iron species cross at 0.771 V while the concentrations and α's cross at 0.059 log 0.444, or 0.021 V less, at 0.750 V.

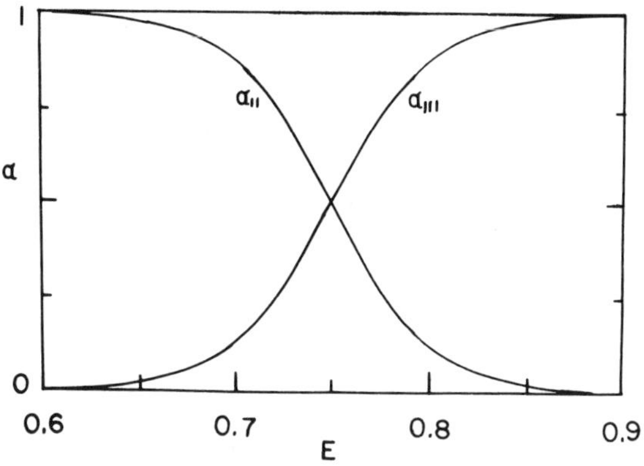

Figure 12-3. The fractions of ferrous (α_{II}) and ferric (α_{III}) versus E in volts for 0.1 M ionic strength.

The logs of the activities have been used in Figure 12-2 to produce the iron diagram. It shows clearly that attempts to obtain cell readings for Fe°–Fe(III) will fail because Fe(II) must first be produced in far higher activity than the Fe(III) to reach equilibrium. It shows that Fe(II) will be lowered below 1/1000 of the Fe(III) if an oxidizing titrant raises the E value of the solution above 1.0 V. It predicts suitable conditions for study of the equilibrium

$$Ag^+ + Fe^{3+} \rightleftharpoons Ag^\circ + Fe^{2+}$$

Namely, we must prepare solutions of the ions comparable to those in Figures 12-1 and 12-2 at 0.6–0.7 V if we are to have all four species present in measurable concentrations. (This is a possible experimental method, although the rate in the direction that dissolves silver metal is extremely slow.) Conversely, the reaction

$$Fe^\circ + 2Ag^+ \rightleftharpoons Fe^{2+} + 2Ag^\circ$$

has no E value that would allow reasonable concentrations of both iron(II) and silver(I) ions. This is the reason that such equilibria are studied via cell potentials, which can be measured at conditions far from equilibrium.

3. Potential versus pH (E–pH) Diagrams

More useful diagrams for many purposes are the E–pH plots giving equilibrium lines between species. These are widely used in geochemical literature.[2,3] Our previous diagrams have not expressed the effects of acidity on redox reactions. None of our diagrams brings in the effects of rate of reaction, which can be limiting (very slow) in redox cases, while rates are commonly fast in acid–base reactions.

To introduce these diagrams, we take our silver case and add the simple pH effect of precipitation of AgOH. (This slowly converts to Ag_2O, but we shall ignore this.) We have

$$Ag^+ + OH^- \rightleftharpoons AgOH_{(c)}, \qquad (Ag^+)(OH^-) = K^\circ_{so} = 10^{-7.71} \quad (12\text{-}8)$$

Taking logs and substituting $10^{-14}/(H^+)$ for (OH^-), we get

$$\log(Ag^+) = 14 - 7.71 - pH = 6.29 - pH \qquad (12\text{-}9)$$

Since our diagrams are to express E and pH, the log activity terms cannot be shown explicitly (Figure 12-4). What is done here is to choose the minimum activity (or concentration in many rough sketches) to use in the equations. For our examples, let us find lines dividing areas having dissolved silver above and below the value 0.1 *M*. Equation (12-9) gives one of these lines. Taking the activity coefficient as 0.75, we get

$$\log(0.075) = -1.125 = 6.29 - pH, \qquad pH = 7.42$$

A vertical line at pH 7.42 expresses the fact that AgOH can precipitate whenever the pH of 0.1 *M* Ag^+ goes above 7.42, independently of E. Another line is given by the Nernst equation (12-3), which shows that if some reducing agent lowers the potential of the solution below 0.733 V, the silver(I) concentration must fall below 0.1 *M* as the metal forms. (Compare Figure 12-1.) We now have a rectangular area inside which silver ion concentration can be 0.1 *M* or greater.

Outside the Ag^+ area, two solids, Ag and AgOH, form. What is the functional relation between them? Since they are in equilibrium, they must have the same value of (Ag^+) in equilibrium with each solid. Thus, we solve equations (12-3) and (12-9) for $\log(Ag^+)$ and set these equal to get

$$E = 1.171 - 0.05916\text{pH} \qquad (12\text{-}10)$$

[2] R. M. Garrels and C. L. Christ, *Solutions, Minerals, and Equilibria*, Harper and Row, New York, 1965, Chapters 7 and 11.

[3] W. Stumm and J. J. Morgan, *Aquatic Chemistry*, Wiley–Interscience, New York, 1970, Chapters 7 and 10.

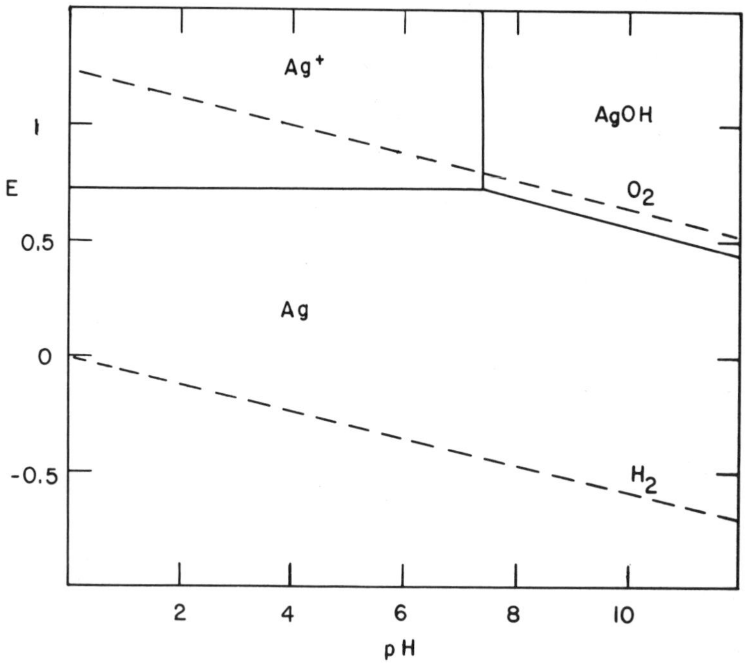

Figure 12-4. *E* in volts versus pH for the silver system. Total dissolved silver species 0.100 *M* at Ag^+ boundary lines.

Interestingly, this is just the Nernst equation for the reaction

$$AgOH + e^- \rightleftharpoons Ag + OH^-$$

It has slope -0.05916, and would extrapolate to 1.171 V at pH zero. This shows the expected dependence: AgOH should be favored as (OH^-) increases. Also, it satisfies both equations (12-3) and (12-9) at pH 7.42 and at 0.733 V, where it meets the other two lines.

This fundamental diagram should be thoroughly understood at this point. The vertical and horizontal lines are boundaries of constant (Ag^+), while the sloping line represents a steadily decreasing (Ag^+) in equilibrium with the two solids. Away from this line only one solid is present. Horizontals must be redox lines, verticals are acid–base changes, while the intermediate slopes show changes involving both acid–base and redox dependence.

One further phenomenon is included in these diagrams: the redox behavior of the water itself. From the half-reactions in Table 12-1,

Table 12-1. Standard Reduction Potentials

Half-reaction	$E^\circ_{1/2}$, V
$Na^+ + e^- \rightleftharpoons Na(s)$	−2.714
$Mn^{2+} + 2e^- \rightleftharpoons Mn(s)$	−1.182
$Fe^{2+} + 2e^- \rightleftharpoons Fe(s)$	−0.440
$2H^+_{aq} + 2e^- \rightleftharpoons H_2(g)$	0.000
$S(s) + 2H^+_{aq} \rightleftharpoons H_2S(g)$	0.141
$AgCl(s) + e^- \rightleftharpoons Ag(s) + Cl^-$	0.222
$Fe^{3+} + e^- \rightleftharpoons Fe^{2+}$	0.771
$MnO_4^{2-} + 2H_2O + 2e^- \rightleftharpoons MnO_2(s) + 4OH^-$	0.51
$Ag^+ + e^- \rightleftharpoons Ag(s)$	0.7991
$O_2(g) + 4H^+_{aq} + 4e^- \rightleftharpoons 2H_2O(l)$	1.223
$MnO_2(s) + 4H^+_{aq} + 2e^- \rightleftharpoons Mn^{2+} + 2H_2O(l)$	1.23
$MnO_4^- + 8H^+_{aq} + 5e^- \rightleftharpoons Mn^{2+} + 4H_2O(l)$	1.51
$MnO_4^- + 4H^+_{aq} + 3e^- \rightleftharpoons MnO_2(s) + 2H_2O(l)$	1.70

the line at which oxygen gas has 1 atm pressure is given by

$$E = 1.223 - 0.05916\text{pH} \tag{12-11}$$

and for hydrogen gas

$$E = 0.000 - 0.05916\text{pH}$$

The two lines for the equilibrium production of hydrogen and oxygen from water are shown dashed in the E–pH diagrams. Above and below them, the water may be decomposed by the oxidant or reductant added. Due to kinetic slowness, high overvoltages of gas production, a significantly wider range of E values is actually attainable in the study of aqueous solutions. Thus, these lines are not sharp limits. There are also approximate limits on the pH attainable. While one can make solutions of pH above 13 and below 1, accurate activity equilibrium predictions cannot be expected for high ionic strength cases (10 M HCl might have pH near −1 and 10 M NaOH near 15).

Very powerful oxidants like Ag^{2+} (2 V) and Co^{3+} (1.8 V) do actually cause oxygen evolution from water, while agents like MnO_4^- (1.5 V), which might be expected to decompose water, do so very slowly at room temperature. Reductants like Na (−2.7 V) rapidly evolve H_2 with water, while Mg (−2.4 V) requires boiling to give noticeable H_2.

Notice that the activity pH, not **pH**, is the natural coordinate for these diagrams.

4. Applications

4.1. E–pH Diagram for Iron

Let us work through in detail a more complex case, that for iron, using ionic strength 0.1 *M*, and 0.01 *M* dissolved species of Fe at the boundary lines. Equations (12-6) and (12-7) give the lines labeled *a* and *b* in Figure 12-5 at −0.51 and +0.771 V. Here we have chosen to use the equal-activity line rather than equal concentrations calculated previously.

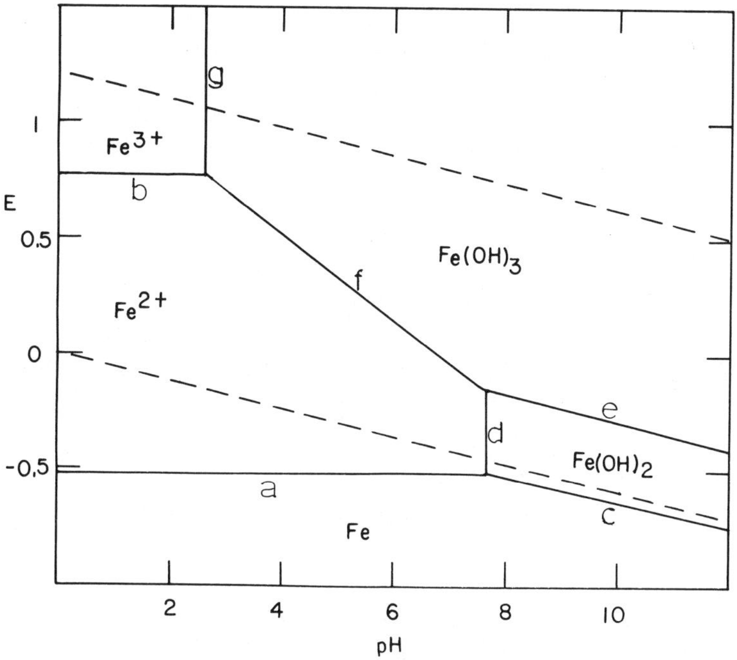

Figure 12-5. E in volts versus pH for the iron system. Dissolved Fe is 0.0100 *M* at the ionic boundary lines. Dashed O_2 and H_2 lines as in Figure 12-4. For *a–g* calculations, see text.

We follow the example of the silver case; the remaining equilibria to be considered and their published E and K values are

(c) $Fe(OH)_2 + 2e^- \rightleftharpoons Fe^\circ + 2OH^-$

$$E = -0.057 - 0.05916\text{pH}$$

(d) $Fe(OH)_2 \rightleftharpoons Fe^{2+} + 2OH$

$$K^\circ_{so} = (Fe^{2+})(OH^-)^2 = 10^{-15.1}$$

(e) $Fe(OH)_3 + e^- \rightleftharpoons Fe(OH)_2 + OH^-$

$$E = 0.305 - 0.05916\text{pH}$$

(f) $Fe(OH)_3 + e^- \rightleftharpoons Fe^{2+} + 3OH^-$

$$E = 1.20 - 0.05916(3)\text{pH}$$

(g) $Fe(OH)_3 \rightleftharpoons Fe^{3+} + 3OH^-$

$$K^\circ_{so} = (Fe^{3+})(OH^-)^3 = 10^{-37}$$

The half-reactions in (c), (e), and (f) are obtained by combining the solubility constant expressions in (d) and (g) with the basic Nernst expressions (a) and (b), equations (12-6) and (12-7). One example should make the method clear. Let us derive the *e* line.

We need the iron(II) to (III) activity ratio for the Nernst expression (12-7) for the situation $Fe(OH)_3$ and $Fe(OH)_2$ both in equilibrium with the ions. We can get this ratio by dividing the K expressions in (d) and (g):

$$\frac{10^{-15.1}}{10^{-37}} = \frac{(Fe^{2+})(OH^-)^2}{(Fe^{3+})(OH^-)^3}$$

Thus, the iron(II) to (III) ratio depends only on (OH^-), in solutions simultaneously saturated with both solids. Putting this ratio into equation (12-7), we obtain the desired E form:

$$E = 0.771 - 0.05916 \log 10^{21.9}(OH^-)$$

Exchanging (OH^-) for its equivalent, $10^{-14}/(H^+)$, and using $\text{pH} = -\log(H^+)$, we convert to

$$E = 0.305 - 0.05916\text{pH}$$

The previous equation in terms of (OH^-) is more correctly the Nernst expression for (e). It is convenient to express all of our equations for the diagram in terms of pH. But, note that the E° of 0.305 V applies to unit activity (H^+) or pH zero. At pH 14, or pOH zero, either form of the equation gives $E = -0.524$ V, which is the $E°$ for the basic form.

From this example and the silver calculations, it should be clear how the rest of the lines in Figure 12-5 were found. We have made several approximations. Where lines b and g meet, the (Fe^{2+}) and the (Fe^{3+}) are both below their maximum values, so that a slight curvature of these lines should exist close to this intersection. Such species as $FeOH^{2+}$, $Fe(OH)_2^+$, Fe_2O_3, and Fe_3O_4 have been ignored. Nevertheless, the diagram will allow interpretations of iron chemistry such as the following:

1. Free iron at zero pH in, say, dilute HCl solution, can give hydrogen and Fe^{2+}. Production of Fe^{3+} is incompatible with the E values required by Fe° and H_2 gas. The reason one gets some iron(III) in practice can be seen on the diagram at the dashed line for O_2, which is present in air at about 0.2 atm partial pressure. At pH zero this produces an oxidizing voltage near 1 V, which puts the solution into the Fe^{3+} region.

2. HNO_3 (E about 1 V) reacts with Fe° to give Fe^{3+} since any Fe(II) formed can easily be oxidized by the nitric acid. The highest E reduction product is not H_2 but either NO or NO_2 depending on pH and concentration.

3. Fe(II) solutions in water are stable in absence of oxygen. But if oxygen is present and no acid has been added, one moves up in the diagram into the $Fe(OH)_3$ region. The Fe^{2+} ion is weakly acidic, so that the original solution may have pH about 5. The E value of, say, 0.1 M $FeSO_4$ is indeterminate. Equation (12-7) shows that some Fe(III) must form (by reaction with water), so that the E is not $-\infty$, as a value of zero for (Fe^{3+}) produces in the Nernst equation. If one part per 10^9 forms Fe(III), the E is about 0.2 V, which lies in the Fe^{2+} stability area at pH 5.

4. Pure $Fe(OH)_2$ cannot be formed in air. The addition of base to precipitate $Fe(OH)_2$ at high pH gives a material which is oxidized at quite a low potential to $Fe(OH)_3$ along line e.

5. Iron is not found exposed at the earth's surface, but has been found in moon rocks. On the diagram Fe° lies below both H_2 and O_2 lines; therefore iron compounds should form at equilibrium (in geological time) if water and O_2 are present to react with it. These seem to be absent, or in very low partial pressure, on the moon.

4.2. E–pH Diagram for Manganese

Let us sketch a rough diagram using zero ionic strength approximations (molarities) and necessarily rough guesses about some $E°$ values for unstable states of manganese. Let us use 0.0100 M of total dissolved species. Table 12-1 and the solubility constant of $Mn(OH)_2$ give us the following equations.

(a) $E = -1.182 + (0.0592/2)\log(Mn^{2+})$ $\quad Mn^{2+} \rightarrow Mn(0)$

(b) $E = 1.23 - (0.0592/2)\log[(Mn^{2+})/(H^+)^4]$

$MnO_2 \rightarrow Mn^{2+}$

At $(Mn^{2+}) = 0.01$, $E = 1.29 - 0.118\text{pH}$.

(c) $E = 0.51 - (0.0592/2)\log[(OH^-)^4/(MnO_4^{2-})]$

$MnO_4^{2-} \rightarrow MnO_2$

At $MnO_4^{2-} = 0.01\ M$, $E = 2.10 - 0.118$ pH.

(d) $E = 1.70 + (0.0592/3)\log[(MnO_4^-)/(H^+)^4]$

$MnO_4^- \rightarrow MnO_2$

(e) $K_{s0} = 10^{-12.8} = (Mn^{2+})(OH^-)^2$

(f) The potentials for Mn^{2+} from Mn(III) and Mn(IV) in solution (not MnO_2) are both near 1.5 V in very acidic solution involving some anion complexing. These regions are only roughly sketched.

Following the same methods as in the previous cases, we derive the diagram in Figure 12-6. A number of facets of manganese chemistry become clear from the diagram:

1. The stable form of manganese at the earth's surface is MnO_2. At normal water pH and the E values due to O_2, the system is in the central MnO_2 region at equilibrium. The Mn^{2+} leached from rocks under water is oxidized to MnO_2 by contact with air. This is a problem in some water supplies. It is removed by addition of lime (CaO) to raise the pH to the region where oxidation to insoluble MnO_2 occurs more rapidly, at a lower potential. Chlorine, which is added anyway, can effect the oxidation more rapidly than O_2.

2. Green manganate(VI) is prepared by heating MnO_2 (lab and industrial method) in molten KOH with either oxygen or chlorate as oxidant. This can be dissolved in water if KOH is present to keep the pH high. (See diagram.) When more water is added, and usually some CO_2, to lower the pH, the solution turns purple and a brown

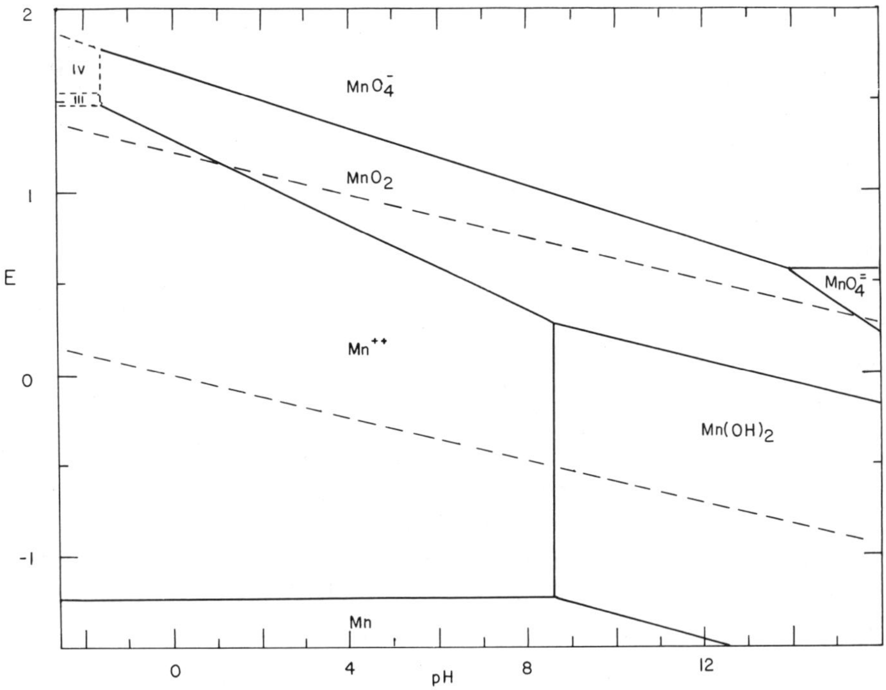

Figure 12-6. *E* in volts versus pH for the manganese system. Total dissolved species is 0.010 *M* at boundary lines of ions. Dashed areas for estimated Mn(III) and Mn(IV) in solution. Dashed O_2 and H_2 lines as in Figure 12-4.

precipitate forms. The diagram explains this. Moving left from the MnO_4^{2-} region, one goes to permanganate and MnO_2. This is the usual means of preparation of $KMnO_4$, a widely useful strong oxidant. It is made from natural MnO_2, pyrolusite.

3. Attempts to oxidize Mn^{2+} to MnO_4^- fail at pH above zero because mainly MnO_2 forms. In the presence of H_3PO_4, which stabilizes Mn(III) and Mn(IV) in solution, a homogeneous path to Mn(VII) is evidently opened. Complexing will move the boundary between Mn(III) and (IV), and MnO_2 to higher pH, allowing oxidation without trapping the Mn in the very insoluble MnO_2. The powerful oxidants IO_4^- and $S_2O_8^{2-}$ are used to make permanganate for analytical purposes. The MnO_4^- to Mn^{2+} potential in Table 12-1 does not represent one reversible step.

4.3. Effects of Complexing and Precipitating Agents

Any reaction lowering the activity of a redox species affects the potential according to the Nernst equation. Let us show this for the silver system shown in Figure 12-4. What shifts occur if a constant activity of 0.00100 *M* Cl^- is present? The K_{so}° expression allows us to say that the maximum activity (Ag^+) is now

$$(Ag^+) = K_{so}^\circ/(Cl^-) = 10^{-9.75}/0.00100 = 10^{-6.75}$$

In equation (12-3), this gives +0.400 V for the new horizontal between the metal and the +1 oxidation state. In the K_{so}° for AgOH we now find that this solid forms only above pH 13.04. The slope of the Ag(0)–AgOH line is unchanged. This produces Figure 12-7.

An almost identical diagram results if we use a system having constant unbound NH_3 activity at 0.1 *M*. The $K_1^\circ K_2^\circ$ expression gives $10^{-6.34}$ for the (Ag^+), taking dissolved Ag(I) as 0.100 *M*.

Other variables of interest might be chosen. For example, let us construct the E–log (Cl^-) diagram for a silver–pH 10 system. Since chloride activity varies, we choose a constant pH. Using the K_{so}°

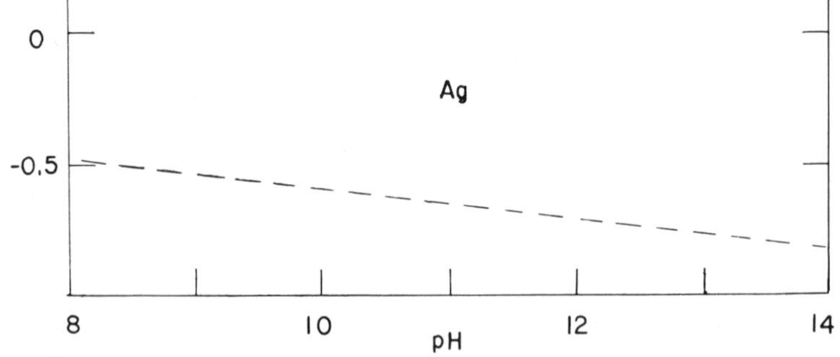

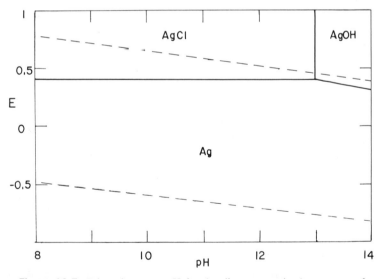

Figure 12-7. E in volts versus pH for the silver system in the presence of a constant 0.00100 *M* activity of chloride ion. Dashed lines for O_2 and H_2 as in Figure 12-4.

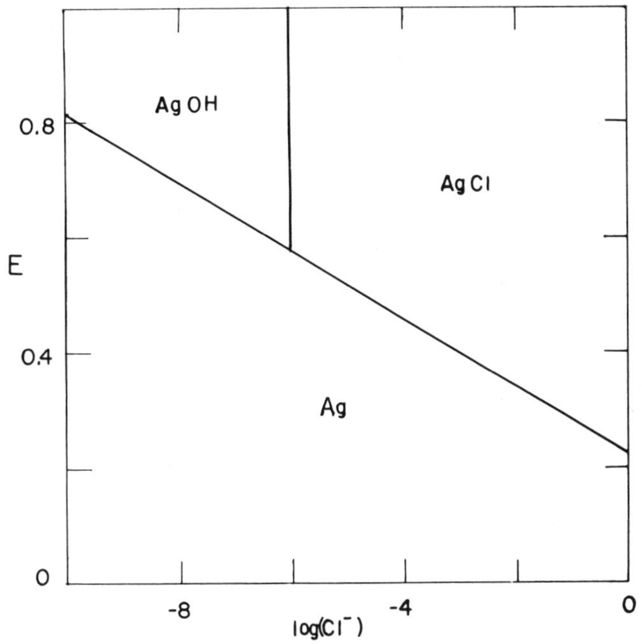

Figure 12-8. *E* in volts versus log (chloride activity) for the silver system at constant pH 10.

above and equation (12-3), we get

$$E = 0.7991 + 0.05916 \log \left[\frac{10^{-9.75}}{(\mathrm{Cl}^-)} \right]$$

$$\mathrm{E} = 0.222 - 0.05916 \log (\mathrm{Cl}^-)$$

Note that this is the E° value for the silver–silver chloride half-cell in Table 12-1. This is the potential for the change from silver(I) to silver(0) in the presence of unit activity chloride. The ratio of the K°_{so} expressions for AgOH and AgCl, $10^{2.04}$, tells us that the two solids are in equilibrium at pH 10 when (Cl^-) is 10^{-6} *M*. The pOH is 4. These relations are plotted in Figure 12-8.

Sulfur and its compounds are of broad occurrence and importance in biochemical and geochemical processes. In preparation for showing sulfur effects on the iron and manganese systems, we show the sulfur system in Figure 12-9. Three acid–base verticals occur: HSO_4^-,

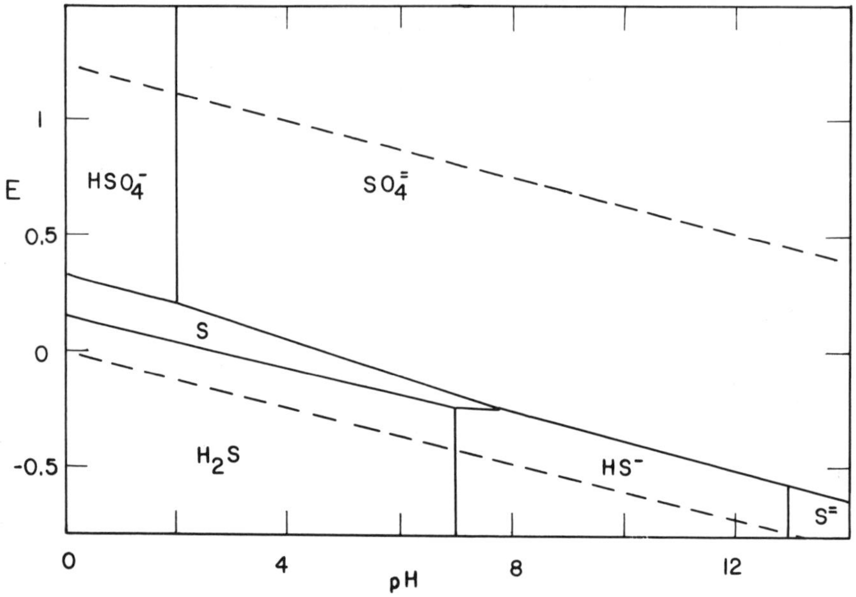

Figure 12-9. E in volts versus pH for the sulfur system. 0.1 M total dissolved species. Adapted from R. M. Garrels and C. L. Christ, *Solutions, Minerals, and Equilibria*, Harper and Row, New York, 1965. Dashed lines for O_2 and H_2 as in Figure 12-4.

pK_a° 1.99; H_2S, pK_1° 7.0 and pK_2° 12.9. It may be surprising that no sulfur(IV) areas appear. SO_2 and the sulfites are not the stable forms under any conditions on the diagram. They are metastable for kinetic reasons and thus known in industrial and laboratory chemistry. In nature, sulfur occurs as sulfates at the surface where O_2 makes high E values, and as sulfur and sulfides under water and deeper in the earth under reducing conditions. When limestones are dissolved in acid, the odor of H_2S can sometimes be noticed. In keeping with the reduced condition, this suggests that iron is found as Fe^{2+} in these freshly made solutions.

Carbonates are also important in sedimentary minerals. On E–pH diagrams only the pK_1 and pK_2 verticals of carbonate need be considered, since no redox of carbon takes place under the conditions present.

In nature, all the relevant systems present must be taken into account in deciding what may be the most stable set of compounds and species on an E-pH diagram. The general method is to find the

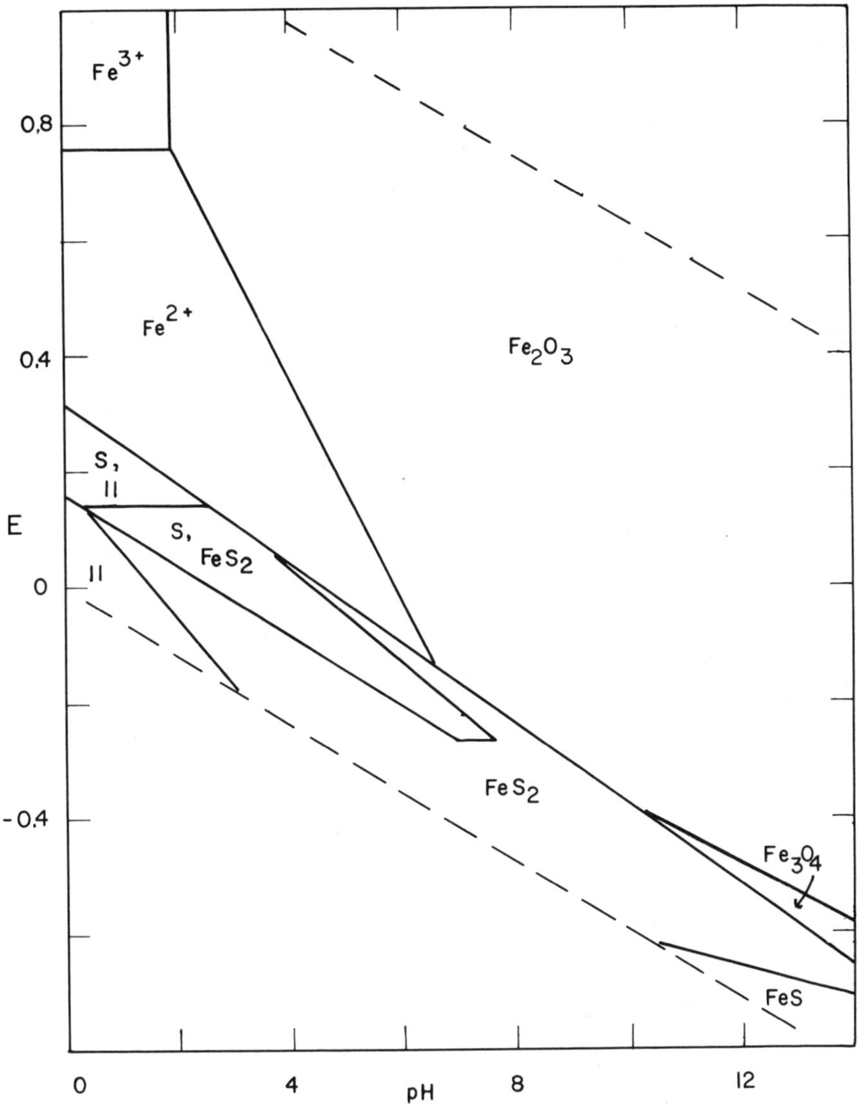

Figure 12-10. E in volts versus pH for the systems iron and sulfur at 25°, 1 atm total pressure. Total dissolved Fe is 10^{-6} M and total dissolved S is 0.1 M at boundaries. From R. M. Garrels and C. L. Christ, *Solutions, Minerals, and Equilibria*, Harper and Row, New York, 1965. Fe^{2+} is abbreviated II in small areas.

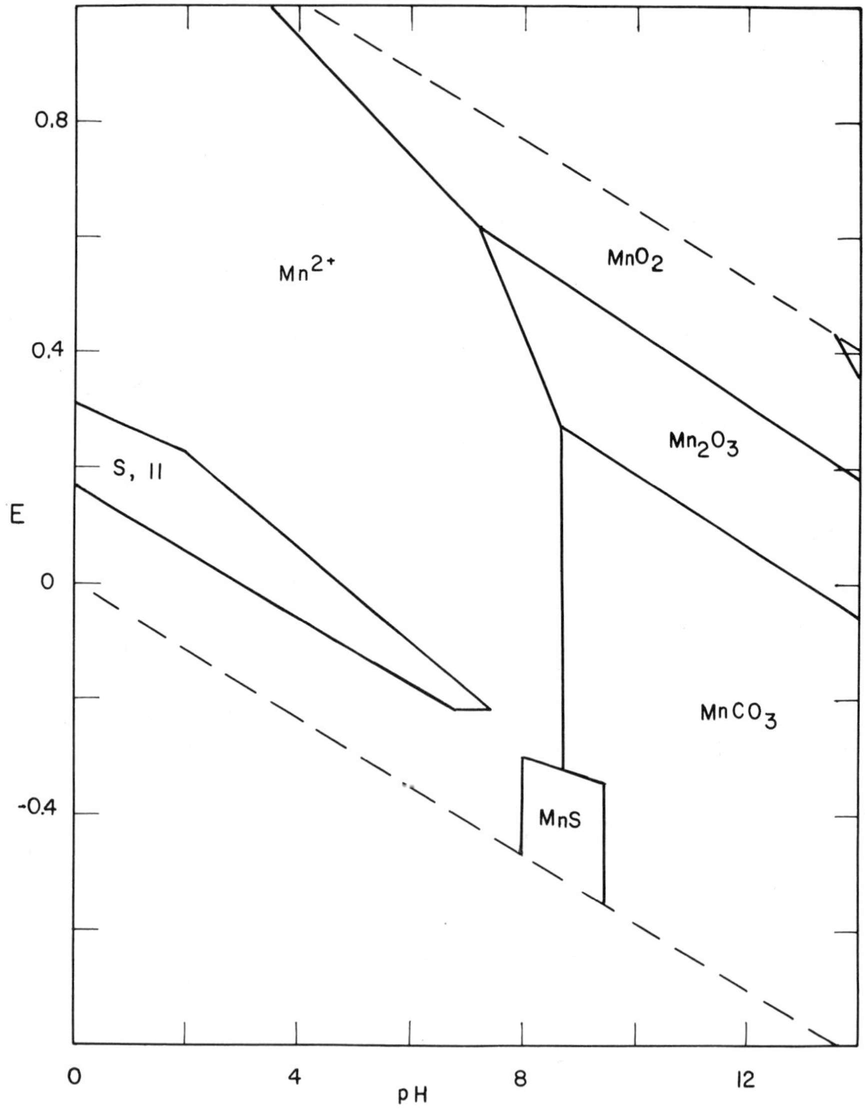

Figure 12-11. E in volts versus pH for the systems Mn, S, H_2CO_3 at 25°, 1 atm total pressure. Total dissolved Mn species at boundaries is 10^{-6} M. Total dissolved S species is 0.1 M. The P_{CO_2} is 10^{-4} atm. Mn^{2+} is abbreviated II in S area. Dashed lines for O_2 and H_2 as in Figure 12-4. Small MnO_4^{2-} area at right: compare Figure 12-6. Species outside water stability region (O_2 and H_2) not shown. From R. M. Garrels and C. L. Christ, *Solutions, Minerals, and Equilibria*, Harper and Row, New York, 1965.

lowest activity species as we have done. The details are complex and only the results of extensive work by geochemists will be shown here (see footnote 2).

Figure 12-10 is a result of combining iron species with likely natural dissolved materials. Hydroxide and sulfide dominate iron chemistry rather than carbonates. This was done for total dissolved iron species at 10^{-6} *M* and total dissolved sulfur species at 0.1 *M*. Initially formed $Fe(OH)_2$ and $Fe(OH)_3$ dehydrate and may combine to form more stable solids, Fe_2O_3 and Fe_3O_4 ($FeO + Fe_2O_3$). With this in mind, one can see the relation of this diagram to our first treatment of Fe in Figure 12-5. The sulfur system brings in the new solids FeS and FeS_2. At high E values, the sulfates of the S system form no insoluble iron compounds.

Figure 12-11 shows that carbonate does play a role in manganese geochemistry. $MnCO_3$ is a mineral of manganese. On our previous diagram (Figure 12-6) no Mn(III) hydroxides or oxides appear, because no equilibrium constants were found to suggest their formation from solution. In general, equilibrium calculations can only reflect known species, which may emphasize the importance of careful investigation of the composition of systems under study to find out what is present.

Application of chemical equilibrium studies is bringing increasing understanding to biological and geochemical phenomena. Much work remains to be done. This is clear in the changes in these diagrams as new data become available.

Problems

1. Iron has a negative electrode potential. Copper is positive, +0.337 V. Iron dissolves in a copper sulfate solution, yet copper dissolves in a ferric chloride solution. Explain, using E values. See Table 12-1.

2. Iron(II) has been used in photographic developing [to reduce Ag(I)]. This works in an oxalate solution. Explain, using E values.

3. If a solution of 0.010 *M* silver nitrate and 0.020 *M* X, a 1:1 complexing agent for silver ion, is prepared, what must be the K_1 formation constant for the soluble complex AgX if the silver electrode potential is to be changed to zero? How can potential measurements be used to find values for formation constants?

4. Silver in contact with air and a solution 0.1 *M* in Cl^- corrodes. Consult Figure 12-7 and write chemical equations for the likely reactions at pH 12 and at pH 4.

5. Sketch a log activity versus E diagram and an E–pH diagram for the nickel system. Take $E°$ for Ni(II) → (0) as -0.23 V, and log K°_{so} for $Ni(OH)_2$ as -16.0.

6. Make a more complete E–pH diagram for nickel (see problem 5) by adding

$$NiO_2 + 4H^+_{aq} + 2e^- \rightleftharpoons Ni^{2+} + 2H_2O, \qquad E^\circ = 1.68 \text{ V}$$

$$Ni(OH)_3 + e^- \rightleftharpoons Ni(OH)_2 + OH^-, \qquad E^\circ = 0.48 \text{ V}$$

---

7. Given the Fe system constants in this chapter, revise the E–pH diagram for the presence of a constant excess of 0.010 M SCN^-. Take K_1 for $FeSCN^{2+}$ formation as 1000. Assume Fe(II) is not complexed.

8. a. What is the potential of a silver wire in contact with a solution of analytical concentrations, Ag(I) 0.010 M and EDTA 0.020 M, at pH 7.00? See Table 9-2 and Figure 9-5.
   b. If NaCl is added to the solution in part a to make its analytical concentration 0.0010 M, what pH adjustment will just prevent the precipitation of AgCl?

9. What is the potential for the Fe(III)–Fe(II) couple in the presence of a slight excess of EDTA if total Fe(III) and Fe(II) are equal and if the pH is high enough so that both are quantitatively complexed by EDTA? See Table 9-2.

---

10. Using the experience of Problems 7 and 9, do further calculations and sketch an E–pH diagram for iron in the presence of a constant excess of 0.0010 M EDTA.

* Appendixes

A-1. Ionic Activity Coefficients from Kielland and by the Davies Equation

For most solution equilibrium calculations, it is more convenient to use ionic activity coefficients than mean coefficients for the electrolyte, even though the latter are more rigorously related to the experimental data. At ionic strengths below 0.1 M, the Kielland table (Table A-1)[1] gives excellent agreement with data. The function of Davies[2] deals rather well with somewhat higher I solutions. For singly charged ions the reliabilities are about $\pm 1\%$ at 0.1 M and $\pm 5\%$ at 0.2 M. For higher charges and higher ionic strengths, the Davies equation may be used as a very approximate guide. For relations to data, see Guenther.[3] Note that an activity coefficient does not correct for ion association into complexes, which is almost always present with ions of multiple charge. The equilibrium constant for the association must be found in addition to the activity coefficients (see Davies' book).

Note that ionic strength is not the concentration of the ion in question for most cases. It must be calculated using equation (2-4) and all the electrolytes in the solution.

Note that the Davies coefficients do level out as ionic strength rises. This is the case for experimental results for many, but not all

[1] J. Kielland, *J. Am. Chem. Soc.* **59**, 1675 (1937).

[2] C. W. Davies, *Ion Association*, Butterworth, London, 1962.

[3] W. B. Guenther, *Quantitative Chemistry*, Addison-Wesley, Reading, Massachusetts, 1968. Chapter 9 compares experimental and calculated values of activity coefficients.

Table A-1. Kielland Table of Ionic Activity Coefficients Arranged by the Sizes of Ions

Charge	Size[a] a	Ions	$I = 0.0005$	0.001	0.0025	0.005	0.01	0.025	0.005	0.1
1	2.5	Rb^+, Cs^+, Ag^+, NH_4^+, Tl^+	0.975	0.964	0.945	0.924	0.898	0.85	0.80	0.75
	3	K^+, Cl^-, Br^-, I^-, CN^-, NO_3^-, NO_2^-, OH^-, F^-, ClO_4^-	0.975	0.964	0.945	0.925	0.899	0.85	0.805	0.755
	4	Na^+, IO_3^-, HCO_3^-, HSO_3^-, $H_2PO_4^-$, ClO_2^-, $C_2H_3O_2^-$	0.975	0.964	0.947	0.928	0.902	0.86	0.82	0.775
	6	Li^+, $C_6H_5COO^-$	0.975	0.965	0.948	0.929	0.907	0.87	0.835	0.80
	9	H^+	0.975	0.967	0.950	0.933	0.914	0.88	0.86	0.83
2	4.5	Pb^{2+}, Hg_2^{2+}, SO_4^{2-}, CrO_4^{2-}, CO_3^{2-}, SO_3^{2-}, $C_2O_4^{2-}$, $S_2O_3^{2-}$, H citrate^{2-}	0.903	0.867	0.805	0.742	0.665	0.55	0.455	0.37
	5	Sr^{2+}, Ba^{2+}, Cd^{2+}, Hg^{2+}, S^{2-}, WO_4^{2-}	0.903	0.868	0.805	0.744	0.67	0.555	0.465	0.38
	6	Ca^{2+}, Cu^{2+}, Zn^{2+}, Sn^{2+}, Mn^{2+}, Fe^{2+}, Ni^{2+}, Co^{2+}, Phthalate^{2-}	0.905	0.870	0.809	0.749	0.675	0.57	0.485	0.405
	8	Mg^{2+}, Be^{2+}	0.906	0.872	0.813	0.755	0.69	0.595	0.52	0.45
3	4	PO_4^{3-}, $Fe(CN)_6^{3-}$, $Cr(NH_3)_6^{3+}$	0.796	0.725	0.612	0.505	0.395	0.25	0.16	0.095
	9	Al^{3+}, Fe^{3+}, Cr^{3+}, Sc^{3+}, In^{3+}, and rare earths	0.802	0.738	0.632	0.54	0.445	0.325	0.245	0.18

[a] Note that these sizes are rounded values for the *effective* size in water solution and are not the size of the simple ions, unhydrated. For a more detailed discussion see the original paper from which these values are taken [J. Kielland, *J. Am. Chem. Soc.* **59**, 1675 (1937)]. Calculated from

$$\log f = \frac{-0.509\, z^2 \sqrt{I}}{1 + 0.328\, a\sqrt{I}}$$

at 25°C, where a is the effective diameter in angstroms. Note I is not M for 2- and 3-charge ions.

electrolytes in the region $I = 0.1$–$1\ M$. However, the specific ions making up the ionic strength can greatly affect these coefficients at these higher ionic strengths (see Guenther's book).

Table A.2. Calculated Ionic Activity Coefficients for 25° by the Davies Equation[a]

I	1	2	3
0.01	0.90	0.66	0.40
0.05	0.82	0.45	0.17
0.10	0.78	0.37	0.11
0.12	0.77	0.36	0.095
0.14	0.76	0.34	0.088
0.16	0.76	0.32	0.081
0.18	0.75	0.32	0.076
0.20	0.75	0.31	0.072
0.30	0.73	0.30	0.062

[a] $\log f = -0.51z^2[\sqrt{I}/(1 + \sqrt{I}) - 0.30I]$, where I is ionic strength and z is the charge on the ion.

A-2. Computer Program for Plotting Alpha Diagrams

In this appendix a plotting program is given for ten values per pH or log [L] decade, which takes any number of constants and plots choice of α, $\log \alpha$, and $\bar{n}$. The program was written by Dr. C. C. Ross, Jr., Mathematics Department, University of the South.

```
      DECEMBER 14, 1974

ALLPLT

1000  REM   PROGRAM TO PLOT ALPHA(N) FOR ANY STEPWISE EQUILIBRIUM
1010  REM             ON AN HP7202A DIGITAL PLOTTER
1020  REM
1030  REM       WRITTEN BY CLAY C. ROSS, JR.    SEPTEMBER 1974
1040  REM
1050  REM               FINAL VERSION: DECEMBER 1974
1060  Q=10
1070  REM             X-INTERPOLATION FUNCTION:
1080  DEF FNP(X)=(0 MAX ABS((LOG(X)-LOG(S))/(LOG(T)-LOG(S)))*9999) MIN 9999
1090  REM             DECADE DETERMINATION FUNCTION:
1100  DEF FNL(X)=INT(-LOG(X)/LOG(10)+.005)
1110  REM             LOG BASE 10:
1120  DEF FNQ(X)=LOG(X)/D1
1130  REM             VERTICAL SCALING FUNCTION:
1140  DEF FNA(X)=(X-V)/(-V)*(X >= V)-(X<V)
1150  REM  -- BOOLEAN EXPRESSIONS ARE USED THROUGHOUT AS SELECTORS
1160  DIM T$[45],M[22,22],A$[3]
```

```
1170  DIM T[2,21],L$[6]
1180  L$="   LOG"
1190  A$=""
1200  N=1
1210  E2=0
1220  V=-10
1230  MAT M=ZER
1240  D1=LOG(10)
1250  PRINT TAB(15);"PLOT OF ALPHA(N)"
1260  PRINT TAB(10);"FOR ANY STEPWISE EQUILIBRIUM"
1270  PRINT LIN(1)
1280  PRINT "ENTER TITLE OF RUN :";
1290  ENTER 255,A,T$
1300  PRINT LIN(2);
1310  PRINT "NUMBER OF K'S";
1320  INPUT E
1330  PRINT "LOG(1) OR ANTILOG(0) FORMAT";
1340  INPUT L
1350  FOR I=1 TO E
1360  PRINT  USING 1370;L$[3*L+1,3*L+3],I
1370  IMAGE #,3A,X,"K(",2D,")="
1380  ENTER 255,J,K[I]
1390  PRINT LIN(1);
1400  IF L=1 THEN 1440
1410  PRINT  USING 1370;L$[4,6],I
1420  PRINT  USING "SDD.DD";LOG(K[I])/D1
1430  GOTO 1470
1440  PRINT  USING 1370;L$[1,3],I
1450  PRINT  USING "D.DDE";10^(K[I])
1460  K[I]=10^(K[I])
1470  PRINT
1480  NEXT I
1490  PRINT "INITIAL AND FINAL A ARE";
1500  INPUT A,F
1510  IF F<A THEN 1540
1520  PRINT "WRONG ORDER: ";
1530  GOTO 1490
1540  A1=A
1550  K=2
1560  REM  E1=1: PLOT NBAR
1570  PRINT "PLOT NBAR";
1580  E1=0
1590  INPUT T$
1600  IF T$[1,1]="N" THEN 1620
1610  E1=1
1620  E3=1
1630  REM  E3=1: PLOT ALPHA CURVES
1640  PRINT "SUPPRESS PLOTTING ALPHA CURVES";
1650  INPUT T$
1660  IF T$[1,1]="N" THEN 1690
1670  E3=0
1680  GOTO 1780
1690  E2=0
1700  REM  E2=1: PLOT LOG(ALPHA)
1710  PRINT "PLOT LOG(ALPHA)";
1720  INPUT T$
1730  IF T$[1,1]="N" THEN 1780
1740  E2=1
1750  PRINT "LOG OF LOWER PLOT LIMIT";
1760  INPUT V
1770  GOTO 1870
1780  PRINT
1790  PRINT "NUMBER OF PLOTS PER PAGE";
1800  INPUT N
1810  IF N=1 THEN 1870
1820  PRINT "THIS IS PLOT NUMBER";
1830  INPUT N1
1840  GOTO 1880
1850  IMAGE 4D,X,+D
1860  IMAGE 4D,X,4D,"^"
```

```
1870  N1=1
1880  D=INT(1/N*9999*.9^(N#1)) MIN 9999
1890  L=INT((N-N1)/N*9999)+100*(N#1)
1900  IF A$#"" THEN 2100
1910  PRINT "PLOT LR OR RL";
1920  INPUT A$
1930  PRINT "RELATIVE PLOT LIMITS:LEFT, THEN RIGHT";
1940  INPUT S,T
1950  F1=1
1960  IF A$[1,2]="RL" THEN 2050
1970  REM  --PROCESS LR PLOT--
1980  F1=0
1990  IF T<S THEN 2020
2000  PRINT "INCONSISTENT WITH ";A$
2010  GOTO 1930
2020  S=S MAX A
2030  T=T MIN F
2040  GOTO 2100
2050  REM  --PROCESS RL PLOT--
2060  IF S<T THEN 2080
2070  GOTO 2000
2080  S=S MIN F
2090  T=T MAX A
2100  PRINT LIN(1);"NUMBER OF AXES (0,2,OR 4)";
2110  INPUT A2
2120  IF A2*(A2-2)*(A2-4)#0 THEN 2100
2130  PRINT "SET PAPER AND PRESS RETURN"
2140  ENTER 255,P,P
2150  PRINT "PLTL"
2160  IF E1=0 AND E3=0 THEN 2590
2170  REM  --DECADE COUNTER LOOP--
2180  FOR C=FNL(A)+1 TO FNL(F)
2190  REM  --INTRA-DECADE PROCESSOR LOOP--
2200  FOR J=1 TO Q+1
2210  MAT T=ZER
2220  REM  --GET LOG OF EACH TERM--
2230  B=0
2240  T[1,1]=FNQ(K[1])+FNQ(A)
2250  B=B MAX T[1,1] MAX 0
2260  IF E=1 THEN 2310
2270  FOR I=2 TO E
2280  T[1,I]=T[1,I-1]+FNQ(K[I])+FNQ(A)
2290  B=B MAX T[1,I] MAX 0
2300  NEXT I
2310  REM  --NORMALIZE EXPONENTS--
2320  A0=(0 <= B AND B<20)*10^((0<B AND B<20)*(-B))
2330  B1=-30 MAX B MIN 30
2340  REM  --RECONSTRUCT SIGNIFICANT NORMALIZED TERMS--
2350  FOR I=1 TO E
2360  IF T[1,I]-B >= V THEN 2390
2370  T[2,I]=-1
2380  GOTO 2410
2390  T[2,I]=10^(T[1,I]-B)
2400  A0=A0+T[2,I]
2410  NEXT I
2420  REM  --GET LOG OF RECONSTRUCTED POLYNOMIAL--
2430  A9=FNQ(A0)
2440  REM  --REBUILD THE ALPHA(N)'S--
2450  M[J,1]=(-A9>V)/A0*(10^(-B1))
2460  IF E2#1 THEN 2480
2470  M[J,1]=FNA(FNQ(M[J,1]))
2480  M[J,21]=0
2490  FOR I=1 TO E
2500  M[J,I+1]=(T[1,I]-A9-B>V)*(T[2,I]#-1)*T[2,I]/A0
2510  M[J,21]=M[J,21]+I*M[J,I+1]
2520  IF E2#1 THEN 2550
2530  M[J,I+1]=(10^(V-1))*(M[J,I+1]=0)+M[J,I+1]*(M[J,I+1]#0)
2540  M[J,I+1]=FNA(FNQ(M[J,I+1]))
2550  NEXT I
2560  M[J,22]=FNP(A)
```

```
2570   A=(10^(-1/Q))*A
2580   NEXT J
2590   REM  --PRE-PLOT CONVERSIONS FOLLOW--
2600   IF E1=0 AND E3=0 THEN 3000
2610   FOR I=1 TO Q+1
2620   FOR K=1 TO E+1
2630   IF I>1 THEN 2650
2640   M[22,K]=0
2650   IF M[I,K] <= 0 THEN 2680
2660   M[22,K]=M[22,K]+M[I,K]
2670   M[I,K]=L+INT(M[I,K]*D)
2680   NEXT K
2690   M[I,21]=L+INT(M[I,21]*D/E)
2700   NEXT I
2710   REM  --OUTPUT IN PROPER PLOT FORMAT--
2720   IF E3=0 THEN 2870
2730   FOR K=1 TO E+1
2740   S1=0
2750   IF E2=1 THEN 2770
2760   IF M[22,K]<.01 THEN 2860
2770   FOR I=1 TO Q+1
2780   IF I>1 AND E2=0 THEN 2840
2790   IF M[I,K]<0 THEN 2850
2800   IF S1=1 THEN 2840
2810   S1=1
2820   PRINT  USING 1860:M[I,22],M[I,K]
2830   GOTO 2850
2840   PRINT  USING 1850:M[I,22],M[I,K]
2850   NEXT I
2860   NEXT K
2870   IF E1=0 THEN 2970
2880   REM  --PLOT NBAR--
2890   PRINT "PLTP"
2900   FOR I=1 TO Q+1
2910   IF I>1 THEN 2940
2920   PRINT  USING 1860:M[I,22],M[I,21]
2930   GOTO 2950
2940   PRINT  USING 1850:M[I,22],M[I,21]
2950   NEXT I
2960   PRINT "PLTL"
2970   REM  --RESET 'A' FOR NEXT PASS--
2980   A=(10^(1/Q))*A
2990   NEXT C
3000   REM  --GENERATE Y-AXIS--
3010   IF E2=1 THEN 3030
3020   K=1
3030   IF A2=0 THEN 3390
3040   A2=A2-2
3050   L2=FNP(F*F1+A1*(1-F1))
3060   PRINT  USING 1860:L2,L+D
3070   B=4+(F-4)*(E1#0)*(K#1)
3080   B=B*(E2=0)-V*(E2=1)
3090   FOR I1=1 TO B
3100   PRINT  USING 1850:L2,L+(B-I1)*D/B
3110   IF I1=B THEN 3190
3120   FOR J1=1 TO 3
3130   IF E1=0 AND K=1 THEN 3160
3140   X=L2+50*(J1=2)*SGN(D)
3150   GOTO 3170
3160   X=L2+((J1=2)*50+(I1=2 AND J1=2)*50)*SGN(D)
3170   PRINT  USING 1850:X,L+(B-I1)*D/B
3180   NEXT J1
3190   NEXT I1
3200   REM  --GENERATE X-AXIS--
3210   FOR I=FNL(A1) TO FNL(F)
3220   L3=FNP(10^(-I))
3230   X=(1-2*(N#1))*(-100 MAX .025*D MIN 100)
3240   IF I#FNL(A1) THEN 3270
3250   PRINT  USING 1860:L3,L
3260   GOTO 3310
```

```
3270  PRINT  USING 1850;L3,L
3280  IF I=FNL(F) AND A2=0 THEN 3310
3290  PRINT  USING 1850;L3,L+X MAX 0
3300  PRINT  USING 1850;L3,L
3310  NEXT I
3320  REM  --CHECK FOR SECONDARY AXES--
3330  IF A2=0 THEN 3390
3340  F1= NOT F1
3350  L=L+D MIN 9999
3360  D=-D
3370  K=2
3380  GOTO 3040
3390  PRINT "PLTT"
3400  F1= NOT F1
3410  E1=E2=0
3420  PRINT "RE-USE THE SAME ALPHA DATA";
3430  INPUT T$
3440  IF T$[1,1]="N" THEN 3470
3450  A=A1
3460  GOTO 1550
3470  END
```

A-3. Typical Run of Computer Program of Appendix A-2

This appendix gives a typical run of the plotting program for all plots: α, $\log \alpha$, and $\bar{n}$. The five arrows indicate five characters corrected in $\log K_1$. The underlines are the operator's answers to requests (Y = yes, N = no).

```
                 PLOT OF ALPHA(N)
           FOR ANY STEPWISE EQUILIBRIUM

ENTER TITLE OF RUN :H3PO4 LOG AND ALL PLOTS
NUMBER OF K'S?3
LOG(1) OR ANTILOG(Ø) FORMAT?1
LOG K( 1)=LL.67←←←←←11.67
    K( 1)=4.68E+11
LOG K( 2)=6.8
    K( 2)=6.31E+Ø6
LOG K( 3)=1.95
    K( 3)=8.91E+Ø1
INITIAL AND FINAL A ARE?1,1E-14
PLOT NBAR?N
SUPPRESS PLOTTING ALPHA CURVES?N
PLOT LOG(ALPHA)?Y
LOG OF LOWER PLOT LIMIT?-12
PLOT LR OR RL?LR
RELATIVE PLOT LIMITS:LEFT, THEN RIGHT?1,1E-14
NUMBER OF AXES (Ø,2,OR 4)?2
SET PAPER AND PRESS RETURN
PLTL
RE-USE THE SAME ALPHA DATA?Y
PLOT NBAR?N
SUPPRESS PLOTTING ALPHA CURVES?N
PLOT LOG(ALPHA)?N
```

```
NUMBER OF PLOTS PER PAGE?3
THIS IS PLOT NUMBER?1
NUMBER OF AXES (Ø,2,or 4)?4
SET PAPER AND PRESS RETURN
PLTL
RE-USE THE SAME ALPHA DATA?Y
PLOT NBAR?Y
SUPPRESS PLOTTING ALPHA CURVES?Y
NUMBER OF PLOTS PER PAGE?2
THIS IS PLOT NUMBER?2
NUMBER OF AXES (Ø,2,OR 4)?4
SET PAPER AND PRESS RETURN
PLTL
RE-USE THE SAME ALPHA DATA?N
DONE
```

A-4. Computer Program for Numerical Values of α and $\bar{n}$

This appendix gives a computer program for numerical calculation of four values of α and $\bar{n}$ per decade of pH or log [L]. Written in UTO FORTRAN in 1969 by Shelburne Wilson and translated to FORTRAN II by Charles Mauthe in 1974. Six constants only.

```
1Ø  REM  PROGRAM FOR ALPHA (N) AND NBAR FOR ANY STEPWISE EQUILIBRIUM
2Ø  REM
3Ø  DIM T$[45]
4Ø  PRINT SPA(2Ø)"ALPHA (N) and NBAR FOR STEP WISE EQUILIBRIUM"LIN(1)
5Ø  PRINT "ENTER TITLE OF RUN :";
6Ø  ENTER 255,A,T$
7Ø  PRINT LIN(1)"ENTER K(1)-K(6)";
8Ø  INPUT K1,K2,K3,K4,K5,K6
9Ø  PRINT "ENTER INITIAL AND FINAL A";
1ØØ  INPUT A,F
11Ø  PRINT "SET PAPER AND PUSH RETURN"
12Ø  ENTER 255,P,P
13Ø  PRINT T$
14Ø  PRINT "K1-K3 "K1;K2;K3
15Ø  PRINT "K4-K6 "K4;K5;K6
16Ø  PRINT
17Ø  FOR J=1 TO 4
18Ø  AØ=1+K1*A*(1+K2*A*(1+K3*A*(1+K4*A*(1+K5*A*(1+K6*A)))))
19Ø  AØ=1/AØ
2ØØ  A1=K1*A*AØ
21Ø  A2=K2*A*A1
22Ø  A3=K3*A*A2
23Ø  A4=K4*A*A3
24Ø  A5=K5*A*A4
25Ø  A6=K6*A*A5
26Ø  B=A1+2*A2+3*A3+4*A4+5*A5+6*A6
27Ø  B=B=.ØØØ5
```

```
280  PRINT LIN(1)"A = "A"     NBAR = "B"     ALPHA (Ø) = "AØ
290  PRINT "ALPHA (1–3) = "A1;A2;A3
300  PRINT "ALPHA (4–6) = "A4;A5;A6
310  A=.562341*A
320  IF A–F<Ø THEN 360
330  NEXT J
340  PRINT LIN(1)
350  GOTO 160
360  END
```

A-5. Equilibrium Constants

Metal ion–ligand formation constants are listed in Tables 8-2, 9-1, and 9-2. Redox potentials are found in Table 12-1.

Below are collected some acid–base constants, pK_n, the negative log of the acidity (dissociation) constant. See Chapter 5 for definitions and forms. First, the pK_n° values are given, followed by the conditional values at 25° and 0.1 *M* ionic strength, unless other conditions are stated.

Acid (protonated form)	pK_1°	pK_2°	pK_3° (25°)	pK_1	pK_2	pK_3
HIO_3	0.78	—	—	—	—	—
HSO_4^-	1.99	—	—	1.60	—	—
HCN	9.21	—	—	9.02	—	—
NH_4^+	9.24	—	—	9.29	—	—
HCOOH (formic)	3.75	—	—	3.56	—	—
$HC_2H_3O_2$ (acetic)	4.756	—	—	4.56	—	—
$HC_2H_2O_2Cl$ (chloroacetic)	2.86	—	—	2.66[a]	—	—
$C_2H_4OHCOOH$ (lactic)	3.858	—	—	3.74[b]	—	—
$CH_3COCOOH$ (pyruvic)	2.49	—	—	—	—	—
H_2S	7.0	12.9	—	6.9	12.6	—
H_2CO_3	6.35	10.33	—	6.16	9.93	—
$C_3H_4OH(COOH)_3$ (citric)	3.13	4.76	6.40 (16)	2.93	4.36	5.74 (16) pK_4
H_3PO_4	2.15	7.21	12.32	1.95	6.80	11.67
H_2Gly^+ (glycine)	2.35	9.78	—	2.43	9.62	—
H_3Glu^+ (glutamic)	—	—	—	2.30	4.28	9.67
H_3Lys^{2+} (lysine)	—	—	—	2.18	9.18	10.72[c]
H_4Y (EDTA)	—	—	—	2.07	2.75	6.24 10.34 pK_4

[a] 1 *M* $NaClO_4$, 20°.
[b] 0.2 *M* KCl, 20°.
[c] 0.01 *M*.

A-6. Bibliography

BASOLO, F. and R. G. PEARSON, *Mechanisms of Inorganic Reactions*, Wiley, New York, 1967. Structure, bonding, and hard–soft acid–base ideas are surveyed in the early chapters.

BATES, R. G., *Determination of pH*, 2nd ed., Wiley, New York, 1973. The authoritative work on pH measurement and standards.

BUTLER, J. N., *Ionic Equilibrium*, Addison-Wesley, Reading, Massachusetts, 1964. A detailed treatment of a variety of ionic reactions of interest to chemists and geologists.

DEAN, J. A., *Chemical Separation Methods*, Van Nostrand Reinhold, New York, 1969. Extensive laboratory directions and practical advice on measurements and operation.

FREISER, H., and Q. FERNANDO, *Ionic Equilibria in Analytical Chemistry*, Wiley, New York, 1963. Calculation methods with applications to laboratory analytical situations.

GUENTHER, W. B., *Quantitative Chemistry*, Addison-Wesley, Reading, Massachusetts, 1968. Elementary equilibrium theory and laboratory projects in determination of equilibrium constants of ionic reactions.

HUHEEY, J. E., *Inorganic Chemistry*, Harper and Row, New York, 1972. Chapter 16 surveys metal ion complexing in biological systems.

JONES, M. M., *Elementary Coordination Chemistry*, Prentice-Hall, Englewood Cliffs, New Jersey, 1964. A thorough survey of metal ion–ligand chemistry. Chapter 8 presents methods of measuring equilibrium constants.

KING, E. J., *Acid–Base Equilibria*, Pergamon Press, London, 1965. A detailed treatment of many types of proton transfer equilibria.

KOLTHOFF, I. M. and P. J. ELVING, eds., *Treatise on Analytical Chemistry*, Wiley, New York, 1959ff. Chapters in Part I by Lee, Sillen, Kolthoff, and Bruckenstein on chemical equilibrium and acid–bases.

LAITINEN, H. A. and W. E. HARRIS, *Chemical Analysis*, 2nd ed. McGraw-Hill, New York, 1975. An advanced text with extensive treatment of equilibria in solution analysis and separation processes.

RICCI, J. E., *Hydrogen Ion Concentration*, Princeton Univ. Press, Princeton, New Jersey, 1952. This is an extensive, rigorous treatment of the mathematics of aqueous acid–base equilibria.

RINGBOM, A., *Complexation in Analytical Chemistry*, Wiley, New York, 1963. Extensive mathematical analysis of equilibrium situations in a variety of analytical processses.

SILLEN, L. G. and A. E. MARTELL, *Stability Constants of Metal-Ion Complexes*, Special Publications No. 17, 1964, and No. 25 (Supplement), 1971, The Chemical Society, London. This is the main source of compiled experimental results for acid–base, solubility, redox, and ion–ligand equilibria.

STUMM, W. and J. J. MORGAN, *Aquatic Chemistry*, Wiley–Interscience, New York, 1970. A detailed treatment of natural systems with many applications and problems.

The following four books give an introduction to solution equilibria using more or less rigorous modern approach:

BARD, A. J., *Chemical Equilibrium*, Harper and Row, New York, 1966.

BLACKBURN, T. R., *Equilibrium*, Holt, Rinehart and Winston, New York. 1969.

BUTLER, J. N., *Solubility and pH Calculations*, Addison-Wesley. Reading, Massachusetts, 1964.

FLECK, G. M., *Equilibria in Solution*, Holt, Rinehart and Winston, New York, 1966.

* Answers and Hints for Selected Problems

Answers to one of each type of problem will usually be given. Numerical answers may not be given to the full number of significant figures allowed by the data. The student should report the proper number of digits justified by the data.

Chapter 1

1. HBr, strong acid (like HCl); KOH, a strong base; KBr, neutral (neither ion is acid or base); KNH_2, a strong base; butyric is a weak acid like formic and acetic; NO_3^- is neutral, while the NH_4^+ is a weak acid, making the solution weakly acidic; acetate is a weak base; Cr_{aq}^{3+} is a weak acid.

2. (a) $PO_4^{3-} + H_2O \rightleftharpoons HPO_4^{2-} + OH^-$.
   (b) $NH_4^+ + OH^- \rightleftharpoons NH_3 + H_2O$.
   (c) $NH_3 + OH^- \rightleftharpoons NH_2^- + H_2O$.
   (d) $H_2SO_4 + HC_2H_3O_2 \rightleftharpoons HSO_4^- + H_2C_2H_3O_2^+$.

3. (a) For the same kernel charge, +6, the size increase of the electron cloud pair and the distance from the kernel decrease the ability to bind a proton going from H_2O to H_2Te.

4. Hint: All the sodium compounds listed have basic anions.

5. See 3(a) and notice that electronegativity (and electron cloud withdrawing power) increases upward from Te to S.

Chapter 2

1. a. $HCOOH + H_2O \rightleftharpoons H_3O^+ + HCOO^-$, $K_a^\circ = (H_{aq}^+)(HCOO^-)/(HCOOH)$. The H on the carbon atom is not acidic in water.
   b. The same as (a), or write as a K_b. Na^+ is inert and *not* shown in either formulation.
   c. Treat like ammonia. See the K_b explanation in the chapter.
   d. $H_2SO_4 \rightarrow H_{aq}^+ + HSO_4^-$ (completely), then $HSO_4^- \rightleftharpoons H_{aq}^+ + SO_4^{2-}$.
2. From Chapter 1 and Figure 1-4, we decide that these are all strong acids and bases in water. (a) 0.0020 M **H** and 0.0020 M NO_3^-; **pH** 2.70. No HNO_3 molecules should be shown, assuming strong acid behavior. The only equilibrium is the H_2O ion formation. (d) 0.003 M Na^+, 0.003 M NH_3, 0.003 M OH^-, **pH** 11.5. The **pH** is set by the strong base, which represses the ion formation by NH_3 to a very small concentration (Le Chatelier shift).
3. Assuming I is small and $f = 1$, we get $\mathbf{H} = 10^{-2.73} = 0.0019\ M$. More correctly, at $I \cong 0.002$, $\mathbf{H} = 10^{-2.71}$. In 0.05 M NaCl, $I \cong 0.05$, we get $\mathbf{H} = 10^{-2.65} = 0.0022\ M$.
4. See data in the chapter. pH 7.17 at 15°.
5. From data in the chapter, at 0.05 M, $\mathbf{H} = 1.2 \times 10^{-7}$; pH $= 7.0_2$.
6. 0.0063 M.
7. With each at 1 mol, there is a 33% increase in the ester. For the second case, there is a 67% yield of the ester.

Chapter 3

1. a. 0.50 M HX, 0.50 M X^-.
   b. C_a is 0.10 M NH_4^+, C_b is 0.30 M NH_3.
   c. C_a is 0.30 M HSO_4^-, C_b is 0.20 M SO_4^{2-}.
2. (a) 4.7, (b) 9.7, (c) 1.8, (d) 13.7.

8. 2.510, using Keilland table values for f, and correcting the C_a and C_b in equation (3-2) with the **H** calculated from the pH (activity).
9. pK_a values: acetic acid, −0.20 log units; NH_4^+ no change (Debye–Hückel equation) or +0.05 unit, Keilland; HSO_4^- −0.39 unit. pK_b: no change, −0.25, and +0.19 unit.
10. Start 2.6 pH, $I = 0.0025$, $f_+ = 0.95$, 0.04 log unit change in pK_a, equation (3-2) is on the borderline of 0.01 log unit change in **H**. At midpoint, corrected **pH** is 3.83, equation (3-2) not needed. At end, I is 0.04 M, corrected activity pH is 8.18.

Chapter 4

3. See Figure 4-5.
   a. For acetic acid one is at low **pH**, where α_1 is large, at moderate C_a values. As C_a gets very small, **pH** must approach neutrality and α_0 becomes predominant (Figure 3-2).
   b. NH_3 is major at high **pH**, where α_0 is larger than α_1 in this system. But note the dilution effect as in (a).
   c. Large α fractions of both acetate and NH_4^+ are compatible in the **pH** region between the pK_a values. Thus, we might expect to be half-way between these pK_a values when these species are of equal concentrations. This is true; see further details in Chapter 6.

5. To vary α's, one may vary the buffer ratio taken, or may add strong acid to move left, or strong base to move right. However, to change α' in Figure 4-7, we are limited to chosen C_a or C_b values for a pure acid solution (left half), or for a pure base solution (right half). Thus, the main factor here is the strength of the acid chosen, since these are general curves independent of acid identity. The stronger the acid, the higher the fraction in ionic form, α'_0. The stronger the base, the more of it is protonated, α'_1, moving to the right. Use numbers in equations (4-2)–(4-6) to help clarify the differences.

Chapter 5

8. **pH** 9.1.

Chapter 6

1. (a) 0.2, 0.1, 2.0, (b) 0.1, 0.1, 1.0, (c) 0, 0.1, 0, (d) 0.025, 0.015, 1.67, (e) 0.02, 0.015, 1.33, (f) 0.025, 0.05, 0.5.

2. (a) acid, **pH** 2. (b) Ampholyte, 3.7. (c) Base, 8.7. (d) A buffer at K_1, 2.7. Rough, unrefined values.

3. All species, including Cl^-, have same analytical C. The $\bar{n}$ is 1.8, **pH** 6.2. Slope of $\bar{n}$ curve allows position of $\bar{n} = 1.8$ to be clearly located.

4. Acid means $\bar{n}$ less than 1.0, some HA^{2-} gives protons to water more than it accepts to give H_2A^-. The pK_2 + pK_3 is less than 14. Neutral means $\bar{n} = 1.00$, $K_1K_2 \cong 10^{-14}$.

7. 10 M and 10^{-7} M.

9. **pH** 3.5, 5.3, 7.5.

15. **pH** about 6.7.

Chapter 9

1. Look at the α values at the log [L] of the analytical ligand concentration given to see if much of the ligand will be removed by reaction with M. These cases illustrate very different possibilities.
   a. α_0 and α_1 are major species fractions. This can leave NH_3 just under 1 M and so is consistent with this position on the α diagram.
   b. Here, in the region of log $[NH_3]$ near 0 (1 M), we estimate $\bar{n}$ between 4 and 5, so that much NH_3 has combined with Co^{2+}. Thus, we must move left. Try $\bar{n}$ about 4 to get $[NH_3]$ about $1 - 0.4 = 0.6$, log $[NH_3]$ about -0.2. This gives the species α's decreasing in the order: $\alpha_4, \alpha_5, \alpha_3, \alpha_6, \alpha_2, \alpha_1, \alpha_0$, with only the last two definitely below 1%.
   c. 0.2 M NH_3 suggests mainly $Cu(NH_3)_4^{2+}$, which is not possible as the major species with only a 2/1 ratio for L/M. The $\bar{n}$ cannot be above 2, and thus $[NH_3]$ must be below about $10^{-3.3}$. The α_2 is the major species, with somewhat less α_1 and α_3, much less α_0 and α_4.
   d. Here α_2 can be the major species without having to leave more than a small fraction of the NH_3 unbound.
   e. About equal $\alpha_1, \alpha_0, \alpha_2$.
   f. In order, $\alpha_4, \alpha_3, \alpha_2$. The $\bar{n}$ is just over 3.
   g. α_1 near 0.1 and α_0 near 0.9 is possible. $[Cl^-]$ near 10^{-8}.

2. a. $\bar{n}'$ intersects $\bar{n}$ in Figure 9-1 at 1.2 at NH_3 0.83 M.
   $\alpha_1 = 0.40$, $\alpha_0 = 0.28$, $\alpha_2 = 0.23$, $\alpha_3 = 0.075$, $\alpha_4 = 0.013$.

3. $10^{5.3}$.

4. Near 10^{-2}.

5. Near 1.

6. Note Figure 9-7 cannot be used for 10%. **pH** about 2.5.

7. **pH** 11, all calcium and magnesium, about 80 ml. No end point, **pH** 7.

8. 0.40 mol.

Chapter 10

1. 9. Under 1.

2. 1.05×10^{-5}. The ionic strength yields $f_\pm$ about 0.37 ± 0.02, leading to a solubility of $(2.8 \pm 0.2) \times 10^{-5}$.

3. 7.5 and 4.23×10^{-3}.

4. $K_{so} = 2.7 \times 10^{-3}$. $K_{so}^\circ = 1.7 \times 10^{-3}$.

5. 0.0755 M.

6. (a) 1.42×10^{-4}. (b) $K_{s_1} = 5 \times 10^{-6}$. $S = 1.47 \times 10^{-4}$. (c) 1.4×10^{-3}.

Chapter 11

1. $S = [K_1 K_{s_0} X(1 + K_2 X)]^{1/2}$, where X is unbound $S_2O_3^{2-}$ concentration.
3. (a) $0.0016\mathbf{H}^2$. (b) $10^{-20.8}/P$. (c) Like equation (11-3) with $K_{s_1} = 0$.
4. 0.6 *M*.
5. About 0.2 *M* HgS dissolves in 2 *M* KI solution, and about 0.1 as much in $S_2O_3^{2-}$ solution.
6. 0.4 g and 180 ml gas.
7. At p**H** 3 about 0.09 *M*.

Chapter 12

1. The Fe(III) → (II) potential, +0.77 V, overcomes the copper 0.337 V to give spontaneous redox reaction. It remains true that Fe(II) does not oxidize copper.
2. The oxalate is a good complexer for Fe(III), which needs only slight lowering in E if Fe(II) is to be a good reducer for Ag(I). The silver is present as insoluble halides, so $E°$ is not the working potential here.
3. Assuming 0.01 *M* excess complexing agent at equilibrium gives K_1 about 10^{13}
4. $4Ag + 2H_2O + O_2 \rightarrow 4AgOH$.
   $4Ag + 4Cl^- + O_2 + 4H^+ \rightarrow 4AgCl + 2H_2O$.
7. The Fe(III), here $FeSCN^{2+}$, area becomes larger as aquo Fe(III) concentration is lowered slightly by SCN^- complexing.
8. (a) +0.56 V ($f = 1$). (b) p**H** about 9.9 ($f = 1$).
9. Find the Fe(II)/Fe(III) ratio from the ratio of the K_f values for their EDTA complexes. Since Fe(III) is much more strongly complexed, this leads, in the Nernst equation, to a new E value of +0.13 V.
10. Quite a bit of change is involved here. α_{0Y} (Figure 9-5) puts curvature into the lines at lower pH. The solid hydroxides do not form until much higher pH values than in Figure 12-5. Verticals can be shown at low pH where $Fe^{2+} = FeY^{2-}$, and where $Fe^{3+} = FeY^-$.

* Index